T0324984

Statistical Mechanics of Cellular Systems and Processes

Cells are complex objects, representing a multitude of structures and processes. In order to understand the organization, interaction and hierarchy of these structures and processes, a quantitative understanding is absolutely critical. Traditionally, statistical mechanics based treatment of biological systems have focused on the molecular level, with larger systems being ignored. This book integrates understanding from the molecular to the cellular and multi-cellular level in a quantitative framework that will benefit a wide audience engaged in biological, biochemical, biophysical and clinical research. It will build new bridges of quantitative understanding that link fundamental physical principles governing cellular structure and function with implications in clinical and biomedical contexts.

MUHAMMAD H. ZAMAN is an Assistant Professor in the Department of Biomedical Engineering and Institute for Theoretical Chemistry at the University of Texas, Austin.

Statistical Mechanics of Cellular Systems and Processes

Edited by

MUHAMMAD H. ZAMAN

University of Texas, Austin

CAMBRIDGE
UNIVERSITY PRESS

CAMBRIDGE
UNIVERSITY PRESS

Shaftesbury Road, Cambridge CB2 8EA, United Kingdom

One Liberty Plaza, 20th Floor, New York, NY 10006, USA

477 Williamstown Road, Port Melbourne, VIC 3207, Australia

314–321, 3rd Floor, Plot 3, Splendor Forum, Jasola District Centre, New Delhi – 110025, India

103 Penang Road, #05–06/07, Visioncrest Commercial, Singapore 238467

Cambridge University Press is part of Cambridge University Press & Assessment,
a department of the University of Cambridge.

We share the University's mission to contribute to society through the pursuit of
education, learning and research at the highest international levels of excellence.

www.cambridge.org
Information on this title: www.cambridge.org/9780521886086

© Cambridge University Press & Assessment 2009

First published 2009

A catalogue record for this publication is available from the British Library

Library of Congress Cataloging-in-Publication data
Statistical mechanics of cellular systems and processes / edited by Muhammad H. Zaman.
   p. ; cm.
Includes bibliographical references and index.
ISBN 978-0-521-88608-6 (hardback)
1. Statistical mechanics.   2. Cell biology.   3. Biomechanics.   I. Zaman, Muhammad H.
II. Title.
[DNLM: 1. Cell Physiological Phenomena.   2. Biomechanics.   3. Cellular Structures.
4. Models, Biological. QU 375 S797 2009]
QP517.T48S73 2009
571.6′29–dc22   2008052570

ISBN   978-0-521-88608-6   Hardback

Contents

The color plates are to be found between pages 84 and 85.

Contributors

Jason K. Cheung
Biological and Sterile Product Development
Schering–Plough Research Institute
Summit, NJ, USA

D. Condorelli
Department of Chemical Sciences Section of Biochemistry and
Molecular Biology
University of Catania
Catania, Italy

Michael W. Deem
Department of Bioengineering and Department of Physics and
Astronomy
Rice University
Houston, TX, USA

Aaron R. Dinner
Department of Chemistry
James Franck Institute and Institute for Biophysical Dynamics
University of Chicago
Chicago, IL, USA

Jeffrey R. Errington
Department of Chemical and Biological Engineering
State University of New York at Buffalo
Buffalo, NY, USA

L. M. Floría
Institute for Biocomputation and Physics of Complex Systems
(BIFI) and Departamento de Física de la Materia Condensada

University of Zaragoza
Zaragoza, Spain

L. B. Freund
Division of Engineering
Brown University
Providence, RI, USA

J. Gómez-Gardeñes
Scuola Superiore di Catania
Catania, Italy
and
Institute for Biocomputation and Physics of Complex
Systems (BIFI)
University of Zaragoza
Zaragoza, Spain

V. Latora
Dipartimento di Fisica e Astronomia
University of Catania
Catania, Italy

Y. Moreno
Institute for Biocomputation and Physics of Complex
Systems (BIFI)
University of Zaragoza
Zaragoza, Spain

Vincent K. Shen
Physical and Chemical Properties Division
National Institute of Standards and Technology
Gaithersburg, MD, USA

R. Sinatra
Scuola Superiore di Catania
Catania, Italy

Jun Sun
Department of Bioengineering and Department of
Physics and Astronomy
Rice University
Houston, TX, USA

Thomas M. Truskett
Department of Chemical Engineering and Institute of
Theoretical Chemistry
University of Texas at Austin
Austin, TX, USA

Jin Wang
Department of Chemistry, Department of Physics and
Department of Applied Mathematics
State University of New York at Stony Brook
Stony Brook, NY, USA
and
State Key Laboratory of Electroanalytical Chemistry
Changchun Institute of Applied Chemistry
Chinese Academy of Sciences
People's Republic of China

Aryeh Warmflash
Department of Physics
James Franck Institute and Institute for Biophysical
Dynamics
University of Chicago
Chicago, IL, USA

Tianyi Yang
Department of Physics
University of Texas at Austin
Austin, TX, USA

Muhammad H. Zaman
Department of Biomedical Engineering and Institute for
Theoretical Chemistry
University of Texas at Austin
Austin, TX, USA

Preface

While the application of statistical thermodynamics to study molecular biophysics such as proteins, nucleic acids, and membranes has been around for decades, only in the last few years, have researchers started to study hierarchical and complex cellular processes with the tools of statistical mechanics. The emergence of statistical mechanics in cellular biophysics has shown incredible promise for understanding a number of cellular processes such as complex structure of cytoskeleton, biological networks, cell adhesion, cell signaling, gene expression, and immunological response to pathogens. Collaboration with experimental researchers has led to results that are now emerging as promising therapeutic targets. These studies have provided a first-principle picture that is critical for developing a fundamental biological understanding of cellular processes and diseases and will be instrumental in developing the next generation of efficient therapeutics.

While there has been a surge in publications on topics in cellular systems using tools of statistical mechanics, no book has appeared in the market on this emerging and intellectually fertile discipline. This book aims to fill this void. The main purpose of the book is to introduce and discuss the various approaches and applications of statistical mechanics in cellular systems, in an effort to bridge the gap between physical sciences, cell biology, and medicine.

The book is organized such that the first few chapters deal with the physical aspects of cells' structure, followed by chapters on physiologically critical processes such as signaling, and concluding with a chapter on immunology and public health.

In preparation for this I am grateful to all contributors for their timely contribution and to all the staff at Cambridge University press, in particular Dr. Katrina Halliday and Alison Evans for their help, support, and guidance throughout this process.

I hope that the readers of this book, students, researchers, and healthcare professionals will find it useful, thought-provoking, and intellectually stimulating and that the sample of chapters in this book will lead to discovery of new avenues of research in the field.

1

Concentration and crowding effects on protein stability from a coarse-grained model

JASON K. CHEUNG, VINCENT K. SHEN, JEFFREY R. ERRINGTON, AND THOMAS M. TRUSKETT

Introduction

Most of what we know about protein folding comes from experiments on polypeptides in dilute solutions [1–4] or from theoretical models of isolated proteins in either explicit or implicit solvent [5–12]. However, neither biological cells nor protein solutions encountered in biopharmaceutical development generally classify as dilute. Instead, they are concentrated or "crowded" with solutes such as proteins, sugars, salts, DNA, and fatty acids [13–15]. How does this crowding affect native-state protein stability? Are all crowding agents created equal? If not, can generic structural or chemical features forecast their effects?

To investigate these and other related questions with computer simulations requires models rich enough to capture three parts of the folding problem: the intrinsic free energy of folding of a protein in solvent, the main structural features of the native and denatured states, and the connection between protein structure and effective protein–protein interactions. The model must also be simple enough to allow for the efficient simulation of hundreds to thousands of foldable protein molecules in solution, which precludes the use of atomistically detailed descriptions of either the proteins or the solvent.

We recently developed a coarse-grained modeling strategy that satisfies these criteria. It is not optimized to describe any specific protein solution. Rather, it is a general tool for understanding experimental trends regarding how concentration or crowding impact the thermodynamic stability of globular proteins.

Statistical Mechanics of Cellular Systems and Processes, ed. Muhammad H. Zaman. Published by Cambridge University Press. © Cambridge University Press 2009.

To date, the approach has been used to study how protein concentration affects the folding transition [16], how solution demixing phase transitions (e.g., liquid–liquid phase separation) couple to protein denaturation [17], and how surface anisotropy of the native proteins relates to their unfolding and self-assembly behaviors in solution [18]. In this chapter, we review the modeling strategy and some key insights it has produced.

Coarse-grained modeling strategy

Intrinsic protein stability

A two-state protein molecule in a pure solvent has a temperature- and pressure-dependent thermodynamic preference for either its native (folded) or denatured (unfolded) form [19–24]. The free energy difference between these states ΔG_f^0 quantifies the driving force for folding in the absence of protein–protein or protein–solute interactions. It also determines the equilibrium probability $(1+\exp[\Delta G_f^0/k_B T])^{-1}$ associated with observing the native state in an infinitely dilute solution, where k_B is the Boltzmann constant and T is the temperature.

Interactions that influence this "intrinsic" stability of the native state include, but are not limited to, intra-protein hydrogen bonding, electrostatics, disulfide bonds, and London–van der Waals interactions, as well as effective forces due to excluded volume, chain conformational entropy, and hydrophobic hydration [25]. Here, we focus exclusively on the last three, since they are relevant not only to protein folding [26–29] but also to other self-assembly processes in aqueous solutions [30–32]. Chain conformational entropy and intra-protein excluded volume interactions favor the more expanded denatured state of a protein, while the ability to bury hydrophobic residues in a largely water-free core favors the compact native fold. Intrinsic stability characterizes how these factors for a protein in the infinitely dilute limit balance at a given temperature and pressure.

The coarse-grained modeling strategy we review here [16] calculates ΔG_f^0 under the assumption that a foldable protein can be represented as a collapsible heteropolymer. The effective inter-segment and segment–solvent interactions of the heteropolymer are chosen to qualitatively reflect the aqueous-phase solubilities of the amino acid residues in the protein sequence [33, 34]. One advantage of heteropolymer collapse (HPC) models is that they derive from independently testable principles of polymer physics and hydration thermodynamics. A second advantage is that their behaviors can often be predicted by approximate analytical theories or elementary numerical techniques, which allow them to be efficiently incorporated into multiscale simulation strategies such as the one discussed here. Although HPC theories are descriptive rather than quantitative in nature, the combination of their simplicity and their ability to reproduce experimental folding trends of globular proteins [34] makes them particularly attractive for use in model calculations.

HPC theories often reflect a balance of structural detail and mathematical complexity [9, 35–37]. In our preliminary studies, we use a basic, physically insightful approach introduced by Dill and co-workers [34, 38]. This theory models each protein of N_r amino acid residues as a heteropolymer of $N_s = N_r/1.4$ hydrophobic and polar segments. As is explained in Ref. [38], the factor of 1.4 enters due to a lattice treatment of the protein in which the chain is partitioned into cubic polymer segments. The amino acids in a globular protein can be represented as occupying cubic volumes with an average edge length of 0.53 nm [38], whereas the separation of α-carbons in an actual protein is about 0.38 nm (0.53/0.38 $\approx$ 1.4). The inputs to the theory include temperature T (and, more generally, pH and ionic strength [35]), the number of residues in the protein sequence N_r, the fraction of those residues that are hydrophobic Φ (e.g., based on their aqueous solubilities [33, 34]), and the free energy per unit $k_B T$ associated with hydrating a hydrophobic polymer segment χ. A simple parameterization for χ is available that captures experimental trends for the temperature-dependent partitioning of hydrophobic molecules between a nonpolar condensed phase and liquid water at ambient pressure [16]. Although in this chapter we focus exclusively on thermal effects, we have previously introduced a statistical mechanical method for extending the parameterization for χ to also account for hydrostatic pressure [39].

To compute ΔG_f^0 using this HPC model, one first constructs an imaginary two-step thermodynamic path that reversibly connects the denatured (D) and native (N) states [34]. In step 1, the denatured state with radius of gyration R_D collapses into a randomly condensed configuration with radius of gyration, R_N. The theory assumes that the fraction of solvent-exposed residues that are hydrophobic in both the denatured and the randomly condensed states is Φ, the sequence hydrophobicity of the protein. In step 2, the native state is formed from the randomly condensed state via residue rearrangement at constant radius of gyration, so that the fractional surface hydrophobicity of the protein changes from Φ to Θ. By independently minimizing the free energies of the native and denatured states in this analysis, HPC theory predicts the values of both R_D/R_N and Θ. The intrinsic free energy of folding is the sum of the contributions from the two steps along the imaginary folding path, $\Delta G_f^0 = \Delta G_1^0 + \Delta G_2^0$. Approximate analytical solutions for this HPC theory describe cases where the hydrophobic residues have uniform [34] or patchy [18] spatial distributions on the protein surface. We discuss below how these solutions can, in turn, be used to infer approximate nondirectional and directional protein–protein interactions, respectively.

Non-directional protein–protein interactions

While intrinsic thermodynamic stability governs whether an isolated protein favors the native or denatured state, protein–protein interactions play a role in stabilizing or destabilizing the native state at finite protein concentrations.

Protein–protein interactions reflect protein structure. Since HPC theories provide only coarse information about structure, the effects we discuss here are the most basic, generally pertaining to how protein size and surface chemistry couple to their interactions. We first examine the case where proteins display a virtually uniform spatial distribution of solvent-exposed hydrophobic residues, so that protein–protein interactions are, to first approximation, isotropic. We also limit our discussion to systems where the driving force of proteins to desolvate their hydrophobic surface residues by burying them into hydrophobic patches on neighboring proteins dominates the attractive part of the effective protein–protein interaction. The repulsive contribution to the inter-protein potential accounts for the volume that each protein statistically excludes from the centers of mass of other protein molecules in the solution. As should be expected, the structural differences between folded and unfolded protein states translate into distinct native–native NN, native–denatured ND, and denatured–denatured DD protein–protein interactions.

HPC theory correctly predicts that denatured proteins generally exclude more volume to other proteins $(R_D > R_N)$ as compared to their native-state counterparts as shown in Fig. 1.1a [34]. Moreover, denatured proteins exhibit a greater fractional surface hydrophobicity than folded molecules $(\Phi > \Theta)$. Mean-field approximations [16, 17] predict that the magnitudes of the average "contact" attractions between two isotropic proteins scale as

$$\varepsilon_{ND} = \frac{N_s \chi(T) \Phi \Theta k_B T}{12} \left(\frac{f_e(\rho_s^*)}{[1 + \rho_s^{*-1/3}]^2} + \frac{f_e(1)}{[1 + \rho_s^{*1/3}]^2} \right) \tag{1.1}$$

$$\varepsilon_{NN} = \frac{N_s \chi(T) f_e(1) \Theta^2 k_B T}{24} \tag{1.2}$$

$$\varepsilon_{DD} = \frac{N_s \chi(T) f_e(\rho_s^*) \Phi^2 k_B T}{24} \tag{1.3}$$

where ρ_s^* is the effective polymer segment density, $f_e(\rho_s^*) = 1 - f_i(\rho_s^*)$ is the fraction of residues in the denatured state that are solvent exposed, and $f_i(\rho_s^*) = [1 - (4\pi\rho_s^*/\{3N_s\})^{1/3}]^3$ is the fraction of residues that are on the interior of the protein. A detailed derivation of the above equations can be found in Refs. [16, 18].

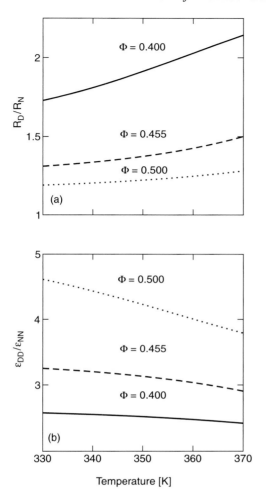

Fig. 1.1 Comparison of the (a) radius of gyration of the denatured protein relative to the native protein, R_D/R_N, and (b) the effective magnitude of the DD attraction relative to the NN attraction, $\varepsilon_{DD}/\varepsilon_{NN}$, for proteins of $N_r = 154$ residues and sequence hydrophobicity $\Phi = 0.400$ (solid), 0.455 (dash), and 0.500 (dot).

Given (1.1–1.3) and $\Phi > \Theta$, it follows that contact attractions involving denatured proteins will generally be stronger than those involving the native state (Fig. 1.1b). This is why denaturation often leads to aggregation and precipitation in protein solutions. Along these lines, attractions between proteins increase in strength with the hydrophobic content of the underlying protein sequence Φ.

In our coarse-grained strategy, the interprotein exclusion diameters, $\sigma_{DD}/\sigma_{NN} \approx R_D/R_N$ and $\sigma_{ND}/\sigma_{NN} \approx (R_N + R_D)/2R_N$, and the contact energies of (1.1–1.3), all of which are derived from HPC theory [34], serve as inputs into an effective protein–protein potential $V_{ij}(r)$ that qualitatively captures many aspects of protein solution thermodynamics and phase behavior (see, e.g., Refs. [40, 41]):

$$V_{ij}(r) = \infty \quad r < \sigma_{ij}$$

$$V_{ij}(r) = \frac{\epsilon_{ij}}{625} \left\{ \frac{1}{[(\frac{r}{\sigma_{ij}})^2 - 1]^6} - \frac{50}{[(\frac{r}{\sigma_{ij}})^2 - 1]^3} \right\} \quad r \geq \sigma_{ij} \tag{1.4}$$

where ij corresponds to the type of interaction NN, ND, or DD.

Directional protein–protein interactions

In Dill and co-workers' original development of this HPC theory, they assume that there are no spatial correlations between solvent-exposed hydrophobic residues in either the denatured or the native state [34]. One way to relax this assumption is to allow for segregation of hydrophobic residues on the surface of the native state. For example, consider the hypothetical scenario where two symmetric "patches" form on the surfaces of native proteins during folding. The patches are distinguishable because they have a different hydrophobic residue composition than the rest of the solvent-exposed "body." As shown in Fig. 1.2, the size of each patch is defined by the polar angle α. The fractional patch hydrophobicity Θ_p and body hydrophobicity Θ_b are expressed as

$$\Theta_p = \frac{f_{ph}\Theta}{1 - \cos\alpha}$$

$$\Theta_b = \frac{(1 - f_{ph})\Theta}{\cos\alpha} \tag{1.5}$$

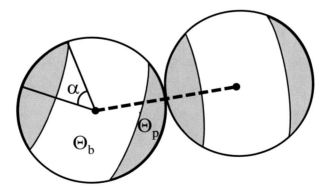

Fig. 1.2 Schematic of two anisotropic native-state proteins. The patch (shaded) and body (white) regions have different hydrophobic residue compositions. The size of the patch is defined by the angle α. The hydrophobicities of the patch Θ_p and body Θ_b are determined by (1.5). Since the dashed line connecting the protein centers passes through a patch region on each molecule, these two proteins are currently in a patch–patch alignment. Adapted from Ref. [18].

where f_{ph} quantifies the fraction of the surface hydrophobic residues that are sequestered into the patch regions on the native protein. Increasing f_{ph} increases Θ_p, which results in higher surface anisotropy and, as we see below, stronger patch–patch attractions. Knowledge of native-state structure would, in principle, allow one to formulate an approximate patch model for a given protein [42], but predicting this structure directly from sequence information using HPC theory is still not generally possible. In other words, f_{ph}, α, and patch location, along with protein sequence, are still knowledge-based inputs for the coarse-grained strategy.

The directional dependencies of the contact attractions of patchy proteins are approximated as follows [18]:

$$\varepsilon_{ND} = \frac{N_s \chi(T) \Phi \Theta_m k_B T}{12} \left(\frac{f_e(\rho_s^*)}{[1 + \rho_s^{*-1/3}]^2} + \frac{f_e(1)}{[1 + \rho_s^{*1/3}]^2} \right) \tag{1.6}$$

$$\varepsilon_{NN} = \frac{N_s \chi(T) f_e(1) \Theta_m \Theta_n k_B T}{24} \tag{1.7}$$

$$\varepsilon_{DD} = \frac{N_s \chi(T) f_e(\rho_s^*) \Phi^2 k_B T}{24}. \tag{1.8}$$

Here, Θ_m and Θ_n denote the apparent surface hydrophobicities associated with different orientational states of interacting native molecules m and n, respectively. For example, to compute the value of Θ_m for molecule m of a given pair interaction, one only needs to know the orientation of molecule m relative to that of the imaginary line connecting its center of mass to that of the other participating protein. If this line passes through a patch on molecule m's surface (see Fig. 1.2), then $\Theta_m = \Theta_p$; otherwise $\Theta_m = \Theta_b$, and so on. Equations (1.6–1.8) reduce to (1.1–1.3) for the isotropic (uniform surface hydrophobicity) case (i.e., $\Theta_p = \Theta_b = \Theta$).

As an illustration, we examine below aqueous solutions of two model proteins of molecular weight $N_s = 110$ (i.e., $N_r = 154$) and hydrophobic residue composition $\Phi = 0.4$, parameters typical for medium-sized, single-domain globular proteins [43]. The difference between the two models is that their native states display distinct surface residue distributions, which in turn lead to different protein–protein interactions: "nondirectional" (i.e., no patches) and "strongly directional" ($f_{ph} = 0.75$, $\alpha = \pi/6$). For simplicity, we refer to these models by their names shown above in quotes, rather than by the f_{ph} and α parameters that define them. As we discuss below, the behaviors of these two model systems provide insights into the mechanisms for stability in several experimental protein solutions.

Figure 1.3 shows the effect of native protein surface anisotropy on the strength of protein–protein attractions. The patch–patch attractions for the

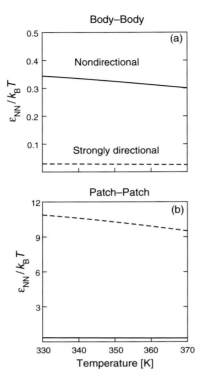

Fig. 1.3 Comparison of the contact attraction ε_{NN} relative to $k_B T$ for (a) body–body alignment and (b) patch–patch alignment of $N_r = 154$, $\Phi = 0.4$ proteins with strongly directional interactions (dash). The contact attraction for $N_r = 154$, $\Phi = 0.4$ native proteins with nondirectional interactions (solid) is also shown.

strongly directional protein are more than an order of magnitude larger than the other attractions. Pairs of directional proteins can desolvate a higher number of hydrophobic residues by self-associating (as compared to the nondirectional proteins), but only if they do so with their hydrophobic patches mutually aligned, which in turn imposes an entropic penalty. This balance between favorable hydrophobic interactions and unfavorable entropy yields the possibility of continuous equilibrium self-assembly transitions involving the native state [18]. Given the symmetric patch geometry of the native state model studied here, the morphology of the self-assembled "clusters" would resemble linear polymeric chains.

The interactions discussed above are similar in spirit to a "two-patch" description that was recently introduced to model the native–native protein interactions of the sickle cell variant of hemoglobin [44] and also to other semi-empirical anisotropic potentials developed for native proteins [42, 45–50].

However, the coarse-grained strategy described here differs significantly from these earlier models in two ways: it explicitly accounts for the possibility of protein denaturation, and it estimates the intrinsic properties of the native and denatured states using a statistical mechanical theory for heteropolymer collapse. This link to the polymeric aspect of the protein is crucial because it allows our model to be used as a tool to investigate how native-state protein anisotropy affects folding equilibria, self-assembly, and the global phase behavior of protein solutions.

Reducing protein stability to a classic chemical engineering problem

It is worth emphasizing that the coarse-grained model described above represents an effective binary mixture of folded and unfolded proteins (the aqueous solvent only entering through χ) connected via the protein folding "reaction." Links between the intrinsic native-state stability of the proteins, ΔG_f^0, the physical parameters defining the protein sequence (N_r, Φ), the native-state surface morphology (Θ, α, f_{ph}), the interactions of hydrophobic residues with aqueous solvent χ, and the protein–protein interactions ($\epsilon_{ij}, \sigma_{ij}$) are established through the HPC model [16, 17]. As in experimental protein solutions, the fraction of proteins in the native state generally depends on both temperature and protein concentration. This fact, often neglected in other modeling approaches which ignore the polymeric nature of proteins or protein–protein interactions, arises because temperature affects the intrinsic stability of the native state ΔG_f^0, and both temperature and protein concentration influence the local structural and energetic environments that native and denatured proteins sample in solution.

In short, the coarse-grained approach frames the stability of concentrated protein solutions in terms of a classic chemical engineering problem: mapping out the equilibrium states of a reactive, phase-separating mixture [51]. In the next section of this chapter, we discuss how advanced Monte Carlo methods designed to efficiently solve the latter problem can also be used to address the former.

Simulation methods

The properties of the coarse-grained protein model described above can be readily studied using transition-matrix Monte Carlo (TMMC) simulation. TMMC is a relatively new simulation technique that has emerged in recent years as a highly efficient method for investigating the thermophysical properties of fluids. It is useful for a variety of applications ranging from the precise calculation of thermodynamic properties of pure and multicomponent fluids in bulk and confinement [52–62], surface tension of pure fluids and mixtures [52, 63–65] and Henry's constants [66] to the investigation of wetting transitions [67, 68] and adsorption isotherms [69, 70]. The basic goal of TMMC is to calculate the free energy of a system along some order parameter path by determining the order parameter probability distribution. To do this in a conventional simulation, a histogram is constructed by simply counting the number of times the system takes on, or visits, a given order parameter value. In a transition-matrix-based approach, the distribution is determined by accumulating the transition probabilities of the system moving from one order parameter value to another during the course of a Monte Carlo simulation, and subsequently applying a detailed balance condition over the explored region of order parameter space. To facilitate sufficient and uniform sampling of order parameter space, a self-adaptive biasing scheme is often employed. The reader is referred to Refs. [52, 56–58, 63, 71–75] for further details. For this coarse-grained protein model, we use a particular TMMC implementation designed for multicomponent systems [57].

To obtain thermodynamic and structural quantities of interest, we perform TMMC simulations within the grand-canonical ensemble. Under these conditions (fixed chemical potentials, volume, and temperature), an appropriate choice of order parameter is the total number of molecules in the system. While the order-parameter probability distribution is unique to the chemical potentials used in the simulation, histogram reweighting can be used to determine the distribution at other chemical potentials [76]. Because the coarse-grained protein model can be regarded as a binary mixture of native and denatured proteins where the components can undergo a unimolecular chemical reaction (folding), chemical equilibrium requires that the chemical potentials of the native and denatured proteins be identical. Thus, only a single chemical potential needs to be specified in a TMMC simulation of the protein solution. The free-energy change of the reaction, that is the intrinsic free energy of folding ΔG_f^0, enters as an activity difference between the folded and unfolded proteins, a treatment which implicitly assumes that the protein's intermolecular and intramolecular degrees of freedom are separable.

Although the main output of a TMMC simulation is the total protein number (concentration) distribution, other system properties can be also calculated. This is done in a straightforward way by collecting isochoric averages during the course of a simulation. Combining these statistics with the above-mentioned histogram-reweighting technique, the fluid-phase properties of the coarse-grained protein model can be determined over a wide range of concentrations from a single TMMC simulation. In our studies, we calculate, along with other properties, the following quantities: the overall fraction of folded proteins f_N, the fraction of clustered or aggregated proteins f_{clust}, and the average fraction of clustered proteins that are folded f_{Nc} [18]. Since this is a binary system, f_N is tantamount to the composition. The midpoint unfolding transition in a protein solution occurs at $f_N = 0.5$. The quantity f_{clust} measures the fraction of proteins in the system that are in clusters. This has also been referred to as the extent of polymerization in other contexts [77]. Here, two proteins are considered to be in the same cluster if the magnitude of their effective pairwise attraction is greater than 80% of the interprotein potential minimum ϵ_{ij}. Finally, the quantity f_{Nc} characterizes the overall average protein cluster composition.

Besides the conventional assortment of Monte Carlo trial moves (e.g., particle displacements, insertions/deletions, and identity changes), we perform so-called aggregation-volume bias trial moves [78–80]. These specialized trial moves are typically required because strong (relative to $k_B T$) inter-protein interactions can otherwise result in persistent bonded configurations which prevent adequate sampling of phase space. Aggregation-volume bias moves circumvent this difficulty by preferentially performing trial protein displacements and protein insertions or deletions in the immediate vicinity, called the "bonding region," of a randomly chosen molecule, thereby promoting the formation and destruction of bonded configurations. Phase space sampling is further enhanced by combining this suite of moves with multiple first-bead trial insertions and configurational-bias Monte Carlo moves [80–82]. Finally, since we are interested in the liquid-state properties of the coarse-grained protein model, a constraint is imposed to prevent the system from crystallizing. This is required because, for the one-component (all native or all denatured) system, the liquid state is metastable with respect to the solid at intermediate and high concentrations [40]. Specifically, we apply a constraint based upon a bond-orientational metric capable of distinguishing between amorphous (liquid) and solid configurations [83–86]. The interested reader is referred to Ref. [17] for further details.

Effects of protein concentration: uniform vs. patchy proteins

In this section, we examine the predictions of our grand-canonical TMMC simulations and coarse-grained modeling strategy with a focus on elucidating how the surface anisotropy of the native state affects denaturation and equilibrium self-assembly behaviors of proteins in solution. In particular, we compare simulated equilibrium unfolding curves (f_N) and self-association measures (f_{clust} and f_{Nc}) for the nondirectional and the strongly directional model proteins described above.

We first explore the physics that govern the stability of the nondirectional protein in solution as a function of its concentration (Fig. 1.4a). The effects can be seen most clearly by simulating along the $\Delta G_f^0 = 0$ isotherm of the protein,

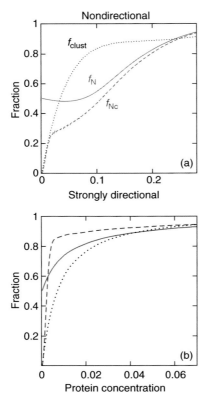

Fig. 1.4 Fraction of folded proteins f_N (solid), "clustered" proteins f_{clust} (dotted), and "clustered" proteins in the native state f_{Nc} (dashed), as a function of protein concentration for the (a) nondirectional and (b) strongly directional model proteins ($N_r = 154$, $\Phi = 0.4$) at their respective infinite-dilution midpoint folding temperatures. Protein concentration is measured as the dimensionless density (i.e., $N\sigma_{NN}^3/V$) where N is the total number of proteins and V is the simulation box volume. To provide a comparison to experimental concentrations, a 100 mg/ml solution of ribonuclease A would have a dimensionless density $N\sigma_{NN}^3/V$ of approximately 0.28, assuming $\sigma_{NN} = 4.0$ nm and a molecular weight of 13.7 kDa. Adapted from Ref. [18].

i.e., its infinite dilution unfolding temperature. As we show below, the existence of patches on a protein serve to modify this baseline behavior.

The minimum in the folded fraction f_N curve for the nondirectional protein in Fig. 1.4a can be understood as a balance between two factors: destabilizing protein–protein attractions involving denatured species and stabilizing protein crowding effects. At finite protein concentrations, a marginally stable protein can favorably denature in solution if it meets two criteria: (a) its current environment affords enough local free volume to accommodate the transition to the more expanded denatured state, and (b) it can form enough new inter-protein hydrophobic contacts with neighboring proteins upon denaturing so that it overcomes its intrinsic free energy penalty for unfolding. The latter destabilizing effect is facilitated by the fact that attractions involving the denatured state are generally much stronger than those involving native nondirectional proteins. Concentrating the protein solution decreases the probability of (a) but increases the likelihood of (b).

To our knowledge, *the coarse-grained strategy outlined here is the first to predict this minimum in folded fraction versus protein concentration*, a nontrivial trend that is observed experimentally in, e.g., monoclonal antibody solutions [87]. Proteins with high hydrophobicity are predicted by the coarse-grained model to show more pronounced concentration-induced destabilization at low and intermediate protein concentrations [16, 17], which also appears to be in agreement with available experimental trends [88].

Does clustering or self-assembly occur in solutions of nondirectional proteins? Figure 1.4a shows that the fraction of clustered nondirectional proteins f_{clust} increases with protein concentration. Clustering of this sort can have two potential origins. On one hand, it can occur for trivial "packing" reasons alone at high protein concentrations, conditions for which pairs of proteins are literally forced to adopt near-contact configurations that satisfy the above geometric criteria for f_{clust}. On the other hand, the more interesting case is when protein clustering arises due to strong inter-protein attractions, which are relevant even at low protein concentrations. For the nondirectional protein discussed above, it is known that interprotein attractions are relatively weak (Fig. 1.3). Thus, clustering for the nondirectional protein is due to geometric "packing" effects. This is also reflected in the fact that the native composition of the clusters f_{Nc} essentially tracks the fraction folded f_N at higher protein concentrations, where most of the proteins are clustered. If, instead, clustering were the result of highly favorable, native-stabilizing inter-protein attractions, then one would expect to find $f_{Nc} > f_N$, even at relatively low protein concentrations.

The behaviors exhibited by the nondirectional protein discussed above are in stark contrast to that of the strongly directional protein presented in Fig. 1.4b. Notice that the relevant protein concentration range for the strongly directional protein is an order of magnitude less than that for the nondirectional proteins. Under these dilute conditions, the packing effects which drive geometric clustering of the proteins are absent. Increases in f_{clust} are therefore a result of different physics: the highly energetically favorable patch–patch association.

The fact that the folded fraction within the clusters f_{Nc} rises above the average folded fraction f_N indicates that the self-assembly behavior or "clustering" stabilizes the patchy native state of the directional protein relative to the denatured state.

In Fig. 1.5, we plot what might be referred to as "stability diagrams" for the (a) nondirectional and (b) strongly directional proteins discussed above. The shaded regions of these diagrams indicate temperature and concentrations that

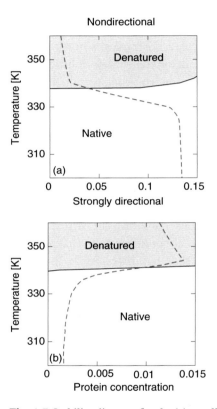

Fig. 1.5 Stability diagram for the (a) nondirectional protein and (b) strongly directional protein in the temperature vs. protein concentration plane. Protein concentration is defined as in Fig. 1.4. The native state is thermodynamically favored ($f_N > 0.5$) in the white region, while the denatured state is favored ($f_N < 0.5$) in the shaded region. Also shown are the loci of conditions where $f_{clust} = 0.5$ (dash). To the right of the $f_{clust} = 0.5$ curve, more than half of the proteins are part of geometric "clusters." For conditions where the denatured state is favored, note that the location of the $f_{clust} = 0.5$ curve for both panels (a) and (b) are the same. This result is expected because the attractive strength and relative size of the denatured proteins are the same for the protein variants studied here. Adapted from Ref. [18].

favor the denatured state ($f_N < 0.5$), while the white region indicates conditions that favor the native state ($f_N > 0.5$). The loci of temperature-dependent concentrations where $f_{clust} = 0.5$ (dash) are also shown. Proteins form clusters in solution for all states to the right of these curves. Note that the clustering behavior of the nondirectional protein occurs at lower protein concentrations for the denatured state as compared to the native state because the former has a larger effective radius of gyration and stronger inter-protein attractions. For the strongly directional proteins, the native-state global clustering trends are different. Small increases in native protein concentration promote self-assembly of linear clusters due to the fact that the patch–patch attractive interactions are much stronger than even the denatured–denatured interactions for this protein.

One key prediction of our model is that strong directional hydrophobic interactions can help stabilize anisotropic native proteins. In other words, folded proteins that associate through hydrophobic patches may show enhanced stability. To illustrate the generality of this result, we briefly discuss below some experimental stability behaviors for proteins in their native states that associate via hydrophobic patches in native or mildly denaturing conditions. We also mention an extreme case where an unstable, unfolded peptide monomer folds only upon specific interaction with other peptide chains.

We begin our discussion with β-lactoglobulin, which exists as a dimer in solution. The interface between the two protein monomers contains a large hydrophobic patch [89]. This geometry suggests that hydrophobic interactions help stabilize the complex. Because the protein must first dissociate to unfold [90], the hydrophobic attractions also aid in preventing nucleation and growth of β-lactoglobulin fibrils [89, 90]. This stability behavior is similar to the enhanced stability seen in the clusters of our strongly directional proteins.

Ribonuclease A is also known to form dimers (as well as trimers and higher-order oligomers) in solution under a variety of experimental conditions [91–93] due to specific, directional interactions at its slightly unfolded N- and C-termini [94]. Like β-lactoglobulin, hydrophobic interactions may help to stabilize the oligomer complex [92]. However, it also has been observed that at high temperatures, the presence of the ribonuclease oligomers is significantly decreased [92]. This oligomer stability behavior is in qualitative agreement with the temperature dependent stability behavior observed for our strongly directional protein clusters. Figure 1.5b shows that the slope of f_{clust} is positive. Therefore, native proteins (unshaded region) that initially favor self-assembly ($f_{clust} > 0.5$) at low temperatures, do not favor self-assembly ($f_{clust} < 0.5$) at higher temperatures.

Sickle-cell hemoglobin has the propensity to form ordered aggregates because of a point mutation that converts a surface hydrophilic residue to a hydrophobic residue [95]. Because wild-type hemoglobin, like our model nondirectional proteins, does not aggregate at low to intermediate concentrations, this protein is an example of how changes to native-state surface hydrophobicity can result in protein polymerization (i.e., the formation of ordered protein clusters). Interestingly, self-association of native sickle-cell hemoglobin can be extrapolated to occur for temperatures above the folding transition of the wild-type protein (see, e.g., Ref. [96]). This suggests that the strongly favorable native interactions of the sickle variant could play an active role in stabilizing the native ("clustering") form over the denatured state.

Finally we review the peptide p53tet. Due in part to its short chain length of 64 residues, the monomeric form of the peptide is unstable. However, when in contact with three neighboring peptide chains, the peptide "folds" to form a tetrameric protein. The interface between the peptides consists of mostly apolar residues [97], suggesting that the hydrophobic interactions can stabilize an inherently unstable protein [97, 98]. In this case, the p53tet tetrameric protein is analogous to our strongly directional model protein clusters, since the clusters are formed by hydrophobic attractions and composed of native proteins (i.e., directional interactions stabilize the geometric formation and the native state).

Does the nature of the crowding species matter?

As mentioned previously, proteins are often a component of crowded solutions in both pharmaceutical and biological settings. For example, the concentration of therapeutic proteins used for subcutaneous injections can be as high as 100 mg/ml [15]. In other cases, proteins are part of a complex mixture of both interacting proteins and weakly interacting (inert) particles. For example, cellular cytoplasm consists of macromolecules, structural proteins (e.g., actin fibrils) and nonstructural proteins [13, 14]. The estimated concentration of the nonstructural proteins is 110 mg/ml [99]. Additionally, inert polymers and macromolecules are often found in pharmaceutical protein solutions [100]. Is the equilibrium stability behavior of proteins in crowded solutions, where there are multiple species present, similar to the equilibrium stability behavior of native proteins in concentrated protein solutions, where there is a single protein species present? In other words, does the type of crowding species strongly affect the equilibrium protein stability?

Thermodynamically, the more compact state (usually the native conformation) is favored in solutions that contain high concentrations of inert crowding species [101–105]. Within the framework of our coarse-grained model, however, we have shown that the stability of native proteins in concentrated protein solution depends not only on stabilizing crowding effects but also on destabilizing protein–protein attractions. The strength of these attractions depends on some of the intrinsic properties of the protein – e.g., the sequence hydrophobicity, the distribution of hydrophobic residues on the native protein surface, and the sizes of the native and denatured states – as well as on solution conditions such as temperature and pressure [16, 17, 39]. For proteins that do not exhibit directional interactions, high sequence hydrophobicity results in non-monotonic concentration effects on stability [16] (see Fig. 1.4a) and liquid–liquid demixing [17] of protein solutions.

To study the effects of crowders on equilibrium protein stability, we extend our original coarse-grained model to account for the presence of multiple species in solution. We compare the equilibrium fraction of folded proteins for two different cases. The first solution consists of foldable proteins (i.e., marginally stable proteins that will fold and unfold) and an ultrastable native crowder (i.e., proteins with interactions identical to the foldable proteins but that remain in their native state and do not unfold). The second solution consists of foldable proteins and inert, hard-sphere crowders with the same effective diameter as the foldable species.

Figure 1.6 shows a preliminary result for a nondirectional protein of chain length $N_r = 154$ and sequence hydrophobicity $\Phi = 0.455$. We observe that the

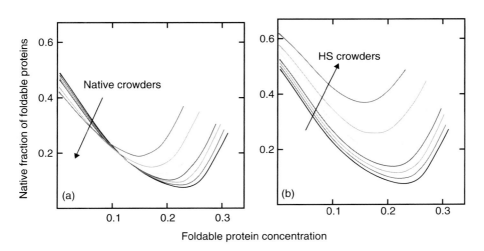

Fig. 1.6 Native fraction of foldable proteins as a function of foldable protein concentration for the protein $N_r = 154$, $\Phi = 0.455$ at its infinite dilution folding temperature. Protein concentration is defined as in Fig. 1.4. The arrow indicates increasing concentration of (a) ultrastable native crowders and (b) inert hard-sphere (HS) crowders. (See plate section for color version.)

ultrastable native crowders can either stabilize or destabilize the native form of the foldable proteins (Fig. 1.6a), depending on the concentration of the latter. For low foldable protein concentrations, increasing the concentration of the native crowders results in a lower fraction of folded proteins. For high foldable protein concentrations, the opposite is true. This nonmonotonic crowding behavior in solution is not observed when hard-sphere crowders are considered, which stabilize the native state at all foldable protein concentrations.

The above calculation illustrates that the magnitude of the crowder–protein attraction plays a nontrivial role. These effects, however, appear to follow from our previously introduced stability criteria. A protein will unfold if (a) there is enough local free volume to accommodate the larger denatured state and (b) denaturing enables formation of sufficiently favorable inter-protein con-tacts. Inert crowders only act to stabilize the native state of the foldable protein because they decrease the likelihood of (a) and do not increase the probability of (b). However, attractive crowders increase the likelihood of (b) at low protein concentrations (where criterion (a) plays a minor role) because denatured–native (and hence denatured–crowder) attractions are larger in magnitude than native–native attractions. However, at high foldable protein concentrations (where criterion (a) dominates), both native or inert crowders reduce the total free volume and thus increase native stability of the foldable protein.

Using multiscale simulations to understand the role of crowder–protein attractions on protein stability is still a relatively new area of inquiry. Progress on this problem, however, is needed as it may help to clarify some of the complicated physics present in heterogeneous protein mixtures like cellu-lar cytoplasm. It may also suggest the physico-chemical properties of new excipients to control the stability of highly concentrated protein therapeutics.

Conclusions and open questions

We have developed a general framework for modeling protein stability in concentrated and crowded solutions. Our approach accounts for both the intrinsic thermodynamics of folding and the general physical characteristics of the native and denatured states. Protein–protein interactions are derived using the salient physical features of the native and denatured conformations predicted by a HPC theory. Ultimately, we are able to study the effects of protein concentration and crowding on protein stability in a computationally efficient manner.

First, we examined aqueous solutions of a single species of nondirectional proteins. At finite concentrations, a marginally stable protein in these systems will unfold if there is enough local free volume and the state-dependent contact free energy of the unfolded protein with its neighbors is sufficiently favorable to overcome any intrinsic thermodynamic stability of the native state. At high protein concentrations, the compact, native conformation is favored over the expanded, denatured state due to self-crowding effects. However, at low concentrations where crowding effects are minimal, a protein may be destabilized due to the more favorable free energies associated with denatured protein–protein contacts. If other factors are equal, we find that proteins with low sequence hydrophobicity tend to be stabilized by increasing protein concentration while proteins with high sequence hydrophobicity show nonmonotonic stability trends [16, 17]. This behavior qualitatively agrees with experimentally observed protein behavior [87, 88].

We also studied the effects of anisotropic protein–protein interactions on protein stability and self-assembly behavior. In contrast with the nondirectional proteins, the strongly directional proteins we studied were stabilized against denaturation at even low protein concentrations by forming highly ordered chains. This behavior is similar to the oligomerization and polymerization of proteins in solution [90, 92, 95].

We are currently examining how different crowding species in solution affect the equilibrium fraction of folded proteins. Our initial results show that attractions play a nontrivial role in determining protein stability. While inert crowders only stabilize marginally stable proteins, attractive crowders can destabilize the native state at low foldable protein concentrations. Here our findings may lead to some insights on the stability of proteins in environments that contain a broad mixture of macromolecules like cellular cytoplasm or concentrated pharmaceutical biological solutions.

Protein solutions show a variety of rich behaviors and open questions remain as to how environmental conditions affect their equilibrium stability trends. For example, one can ask whether the self-assembly of strongly directional proteins will change in the presence of other crowding species. This type of heterogeneous protein solution may more accurately represent in vivo biological environments. One can also ask whether protein solution stability (e.g., the demixing transition observed in Ref. [17]) is affected by protein and inert crowding species. If the equilibrium demixing transition can be prevented by the addition of stabilizing cosolute crowders, protein drug shelf-life may be improved. The general versatility of the coarse-grained framework reviewed here provides the flexibility to investigate many different types of protein solutions commonly encountered in biological and pharmaceutical environments, which should allow it to provide insights into some of these open questions.

Acknowledgments

TMT acknowledges the financial support of the National Science Foundation Grant No. CTS-0448721, the David and Lucile Packard Foundation, and the Alfred P. Sloan Foundation. JRE acknowledges the support of the James D. Watson Investigator Program of the New York State Office of Science, Technology and Academic Research. This study utilized the high-performance computational capabilities of the Biowulf PC/Linux cluster at the National Institutes of Health, Bethesda, MD (http://biowulf.nih.gov) and the Texas Advanced Computing Center (TACC).

References

1. Privalov, P. L., 1979. Stability of proteins: small globular proteins. *Adv. Protein Chem.* **33**:167–241.
2. Schellman, J. A., 1987. The thermodynamic stability of proteins. *Ann. Rev. Biophys. Biophys. Chem.* **16**:115–137.
3. Privalov, P. L., and S. J. Gill, 1988. Stability of protein structure and hydrophobic interaction. *Adv. Prot. Chem.* **39**:191–234.
4. Chalikian, T. V., 2003. Volumetric properties of proteins. *Ann. Rev. Biophys. Biomol. Struct.* **32**:207–235.
5. Dill, K. A., S. Bromberg, K. Z. Yue, *et al.* 1995. Principles of protein folding: a perspective from simple exact models. *Prot. Sci.* **4**:561–602.
6. Shea, J., and C. L. Brooks, 2001. From folding theories to folding proteins: a review and assessment of simulation studies of protein folding and unfolding. *Ann. Rev. Phys. Chem.* **52**:499–535.
7. Mirny, L., and E. Shakhnovich, 2001. Protein folding theory: from lattice to all atom models. *Ann. Rev. Biophys. Biomol. Struct.* **30**:361–396.
8. Snow, C. D., H. Nguyen, V. S. Pande, and M. Gruebele, 2002. Absolute comparison of simulated and experimental protein-folding dynamics. *Nature* **420**:102–106.
9. Pande, V. S., I. Baker, J. Chapman, *et al.* 2002. Atomistic protein folding simulations on the submillisecond time scale using worldwide distributed computing. *Biopolymers* **68**:91–109.
10. Garcia, A. E., and J. N. Onuchic, 2003. Folding a protein in a computer: an atomic description of the folding/unfolding of protein A. *Proc. Natl Acad. Sci. USA* **100**:13 898–13 903.
11. Herges, T., and W. Wenzel, 2004. An all-atom force field for tertiary structure prediction of helical proteins. *Biophys. J.* **87**:3100–3109.
12. Chu, J. W., S. Izvekov, and G. A. Voth, 2006. The multiscale challenge for biomolecular systems: coarse-grained modeling. *Mol. Sim.* **32**:211–218.
13. Goodsell, D. S., 1991. Inside a living cell. *Trends Biochem. Sci.* **16**:203–206.
14. Luby-Phelps, K., 2000. Cytoarchitecture and physical properties of cytoplasm: Volume viscosity, diffusion, intracellular surface area. *Int. Rev. Cytol.* **192**: 189–221.

15. Shire, S. J., Z. Shahrokh, and J. Liu, 2004. Challenges in the development of high protein concentration formulations. *J. Pharm. Sci.* **93**:1390–1402.

16. Cheung, J. K., and T. M. Truskett, 2005. Coarse-grained strategy for modeling protein stability in concentrated solutions. *Biophys. J.* **89**:2372–2384.

17. Shen, V. K., J. K. Cheung, J. R. Errington, and T. M. Truskett, 2006. Coarse-grained strategy for modeling protein stability in solution II: Phase behavior. *Biophys. J.* **90**:1949–1960.

18. Cheung, J. K., V. K. Shen, J. R. Errington, and T. M. Truskett, 2007. Coarse-grained strategy for modeling protein stability in solution III: Directional protein interactions. *Biophys. J.* **92**:4316–4324.

19. Hawley, S. A., 1971. Reversible pressure–temperature denaturation of chymotrypsinogen. *Biochemistry* **10**:2436–2442.

20. Privalov, P. L., 1990. Cold denaturation of proteins. *Crit. Rev. Biochem. Mol. Biol.* **25**:281–305.

21. Smeller, L., 2002. Pressure–temperature phase diagrams of biomolecules. *Biochim. Biophys. Acta* **1595**:11–29.

22. Meersman, F., L. Smeller, and K. Heremans, 2002. Comparative fourier transform infrared spectroscopy study of cold-, pressure-, and heat-induced unfolding and aggregation of myoglobin. *Biophys. J.* **82**:2635–2644.

23. Silva, J. L., D. Foguel, and C. A. Royer, 2001. Pressure provides new insights into protein folding, dynamics, and structure. *Trends Biochem. Sci.* **26**:612–618.

24. Scharngl, C., M. Reif, and J. Friedrich, 2005. Stability of proteins: temperature, pressure, and the role of solvent. *Biochim. Biophys. Acta* **1749**:187–213.

25. Dill, K. A., 1990. Dominant forces in protein folding. *Biochemistry* **29**:7133–7155.

26. Kauzmann, W., 1959. Some factors in the interpretation of protein denaturation. *Adv. Prot. Chem.* **14**:1–63.

27. Tanford, C., 1962. Contribution of hydrophobic interactions to the stability of the globular conformation of proteins. *J. Am. Chem. Soc.* **84**:4240–4247.

28. Meirovitch, H., and H. A. Scheraga, 1980. Empirical studies of hydrophobicity II: Distribution of the hydrophobic, hydrophilic, neutral, and ambivalent amino acids in the interior and exterior layers of native proteins. *Macromolecules* **13**: 1406–1414.

29. Schellman, J. A., 1997. Temperature, stability, and the hydrophobic interaction. *Biophys. J.* **73**:2960–2964.

30. Curtis, R. A., C. Steinbrecher, A. Heinemann, H. W. Blanch, and J. M. Prausnitz, 2002. Hydrophobic forces between protein molecules in aqueous solutions of concentrated electrolyte. *Biophys. Chem.* **98**:249–265.

31. Calamai, M., N. Taddei, M. Stefani, G. Ramponi, and F. Chiti, 2003. Relative influence of hydrophobicity and net charge in the aggregation of two homologous proteins. *Biochemistry* **42**:15 078–15 083.

32. Moelbert, S., B. Normand, and P. De Los Rios, 2004. Solvent-induced micelle formation in a hydrophobic interaction model. *Phys. Rev. E* **69**:061 924/1–061 924/11.

33. Nozaki, Y., and C. Tanford, 1971. The solubility of amino acids and two glycine peptides in aqueous ethanol and dioxane solutions: establishment of a hydrophobicity scale. *J. Biol. Chem.* **1971**:2211–2217.

34. Dill, K. A., D. O. V. Alonso, and K. Hutchinson, 1989. Thermal stability of globular proteins. *Biochemistry* **28**:5439–5449.

35. Alonso, D. O. V., and K. A. Dill, 1991. Solvent denaturation and stabilization of globular proteins. *Biochemistry* **20**:5974–5985.

36. Wallace, D. G., and K. A. Dill, 1996. Treating sequence dependence of protein stability in a mean-field model. *Biopolymers* **39**:115–127.

37. Pande, V. S., A. Y. Grosberg, and T. Tanaka, 2000. Heteropolymer freezing and design: towards physical models of protein folding. *Rev. Mod. Phys.* **72**: 259–314.

38. Dill, K. A., 1985. Theory for the folding and stability of globular proteins. *Biochemistry* **24**:1501–1509.

39. Cheung, J. K., P. Shah, and T. M. Truskett, 2006. Heteropolymer collapse theory for protein folding in the pressure–temperature plane. *Biophys. J.* **91**: 2427–2435.

40. ten Wolde, P. R., and D. Frenkel, 1997. Enhancement of protein crystal nucleation by critical density fluctuations. *Science* **277**:1975–1978.

41. Petsev, D. N., X. Wu, O. Galkin, and P. G. Vekilov, 2003. Thermodynamic functions of concentrated protein solutions from phase equilibria. *J. Phys. Chem. B* **107**:3921–3926.

42. Hloucha, M., J. F. M. Lodge, A. M. Lenhoff, and S. I. Sandler, 2001. A patch–antipatch representation of specific protein interactions. *J. Crystal Growth* **232**:195–203.

43. Shen, M., F. Davis, and A. Sali, 2005. The optimal size of a globular protein domain: A simple sphere-packing model. *Chem. Phys. Lett.* **405**:224–228.

44. Shiryayev, A., X. Li, and J. D. Gunton, 2006. Simple model of sickle hemoglobin. *J. Chem. Phys.* **125**:024 902/1–024 902/8.

45. Sear, R. P., 1999. Phase behavior of a simple model of globular proteins. *J. Chem. Phys.* **111**:4800–4806.

46. Lomakin, A., N. Asherie, and G. B. M. Benedek, 1999. Aeolotopic interactions of globular proteins. *Proc. Natl Acad. Sci. USA* **96**:9465–9468.

47. Curtis, R. A., H. W. Blanch, and J. M. Prausnitz, 2001. Calculation of phase diagrams for aqueous protein solutions. *J. Phys. Chem. B* **105**:2445–2452.

48. Song, X., 2002. Role of anisotropic interactions in protein crystallization. *Phys. Rev. E* **66**:011 909/1–011 909/4.

49. Dixit, N. M., and C. F. Zukoski, 2002. Crystal nucleation rates for particles experiencing anisotropic interactions. *J. Chem. Phys.* **117**:8540–8550.

50. Kern, N., and D. Frenkel, 2003. Fluid–fluid coexistence in colloidal systems with short-ranged strongly directional attraction. *J. Chem. Phys.* **118**:9882–9889.

51. Sandler, S. I., 1999. *Chemical and Engineering Thermodynamics*, 3rd edn. New York: John Wiley.

52. Errington, J. R., 2003. Direct calculations of liquid–vapor phase equilibria from transition matrix Monte Carlo simulations. *J. Chem. Phys.* **118**:9915–9925.

53. Singh, J. K., D. A. Kofke, and J. R. Errington, 2003. Surface tension and vapor–liquid phase coexistence of the square-well fluid. *J. Chem. Phys.* **119**:3405–3412.

54. Shen, V. K., and J. R. Errington, 2004. Metastability and instability in the Lennard–Jones investigated via transition-matrix Monte Carlo. *J. Phys. Chem. B* **108**: 19 595–19 606.

55. Singh, J. K., and D. A. Kofke, 2004. Molecular simulation study of effect of molecular association on vapor–liquid interfacial properties. *J. Chem. Phys.* **121**:9574–9580.

56. Shen, V. K., and J. R. Errington, 2005. Determination of fluid-phase behavior using transition-matrix Monte Carlo: binary Lennard–Jones mixtures. *J. Chem. Phys.* **122**:064 508/1–064 508/17.

57. Errington, J. R., and V. K. Shen, 2005. Direct evaluation of multi-component phase equilibria using flat histogram methods. *J. Chem. Phys.* **123**:164 103/1–164 103/9.

58. Singh, J. K., and J. R. Errington, 2006. Calculation of phase coexistence properties and surface tensions of *n*-alkanes with grand-canonical transition matrix Monte Carlo simulation and finite-size scaling. *J. Phys. Chem. B* **110**:1369–1376.

59. Singh, J. K., J. Adhikari, and S. K. Kwak, 2006. Vapor–liquid phase coexistence for Morse fluids. *Fluid Phase Equilibria* **248**:1–6.

60. Errington, J. R., T. M. Truskett, and J. Mittal, 2006. Excess-entropy-based anomalies for a waterlike fluid. *J. Chem. Phys.* **125**:244 502/1–244 502/8.

61. Mittal, J., J. R. Errington, and T. M. Truskett, 2006. Thermodynamics predicts how confinement modifies the dynamics of the equilibrium hard-sphere fluid. *Phys. Rev. Lett.* **96**:177 804/1–177 804/4.

62. Macdowell, L. G., V. K. Shen, and J. R. Errington, 2006. Nucleation and caviation in spherical, cylindrical, and slablike droplets and bubbles in small systems. *J. Chem. Phys.* **125**:034 705/1–034 705/15.

63. Errington, J. R., 2003. Evaluating surface tension using grand-canonical transition-matrix Monte Carlo simulation and finite-size scaling. *Phys. Rev. E* **67**:012102.

64. Shen, V. K., and J. R. Errington, 2006. Determination of surface tension in binary mixtures using transition-matrix Monte Carlo. *J. Chem. Phys.* **124**:024 721/1–024 721/9.

65. Shen, V., R. D. Mountain, and J. R. Errington, 2007. Comparative study of the effect of tail corrections on surface tension determined by molecular simulation. *J. Phys. Chem. B* **111**:6198–6207.

66. Cichowski, E. C., T. R. Schmidt, and J. R. Errington, 2005. Determination of Henry's law constants through transition matrix Monte Carlo simulation. *Fluid Phase Equilibria* **236**:58–65.

67. Errington, J. R., and D. W. Wilbert, 2005. Prewetting boundary tensions from Monte Carlo simulations. *Phys. Rev. Lett.* **95**:226 107/1–226 107/4.

68. Errington, J. R., 2004. Prewetting transitions for a model argon on solid carbon dioxide system. *Langmuir* **20**:3798–3804.

69. Chen, H., and D. S. Sholl, 2006. Efficient simulation of binary adsorption isotherms using transition matrix Monte Carlo. *Langmuir* **22**:709–716.

70. Chen, H., and D. S. Sholl, 2007. Examining the accuracy of ideal adsorbed solution theory without curve-fitting using transition matrix Monte Carlo simulations. *Langmuir* **23**:6431–6437.

71. Smith, G. R., and A. D. Bruce, 1995. A study of the multi-canonical Monte Carlo method. *J. Phys. A: Math. Gen.* **28**:6623–6643.

72. Wang, J.-S., T. K. Tay, and R. H. Swendsen, 1999. Transition matrix Monte Carlo reweighting dynamics. *Phys. Rev. Lett.* **82**:476–479.

73. Wang, J.-S., and R. H. Swendsen, 2002. Transition matrix Monte Carlo method. *J. Stat. Phys.* **106**:245–285.

74. Fitzgerald, M., R. R. Picard, and R. N. Silver, 1999. Canonical transition probabilities for adaptive metropolis simulation. *Europhys. Lett.* **46**:282–287.

75. Fitzgerald, M., R. R. Picard, and R. N. Silver, 2000. Monte Carlo transition dynamics and variance reduction. *J. Stat. Phys.* **98**:321–345.

76. Ferrenberg, A. M., and R. H. Swendsen, 1988. New Monte Carlo technique for studying phase transitions. *Phys. Rev. Lett.* **61**:2635—2638.

77. Van Workum, K., and J. F. Douglas, 2005. Equilibrium polymerization in the Stockmayer fluid as a model of supermolecular self-organization. *Phys. Rev. E* **71**:031 502/1–031 502/15.

78. Chen, B., and J. I. Siepmann, 2000. A novel Monte Carlo algorithm for simulating strongly associate fluids: applications for water, hydrogen fluoride, and acetic acid. *J. Phys. Chem. B* **104**:8725–8734.

79. Chen, B., and J. I. Siepmann, 2001. Improving the efficiency of the aggregation-volume-bias Monte Carlo. *J. Phys. Chem. B* **105**:11 275–11 282.

80. Chen, B., J. I. Siepmann, K. J. Oh, and M. L. Klein, 2001. Aggregation-volume-bias Monte Carlo simulations of vapor–liquid nucleation barriers for Lennard–Jonesium. *J. Chem. Phys.* **115**:10 903–10 913.

81. Vlugt, T. J. H., M. G. Martin, B. Smit, J. I. Siepmann, and R. Krishna, 1998. Improving the efficiency of the configurational-bias Monte Carlo algorithm. *Mol. Phys.* **94**:727–733.

82. Frenkel, D., and B. Smit, 2002. *Understanding Molecular Simulations: From Algorithms to Applications*, 2nd edn. London: Academic Press.

83. Steinhardt, P. J., D. R. Nelson, and M. Ronchetti, 1983. Bond-orientational order in liquids and glasses. *Phys. Rev. B* **28**:784–805.

84. Torquato, S., T. M. Truskett, and P. G. Debenedetti, 2000. Is random close packing of sphere well defined? *Phys. Rev. Lett.* **84**:2064–2067.

85. Truskett, T. M., S. Torquato, and P. G. Debenedetti, 2000. Towards a quantification of disorder in materials: distinguishing equilibrium and glassy sphere packings. *Phys. Rev. E* **62**:993–1001.

86. Errington, J. R., P. G. Debenedetti, and S. Torquato, 2003. Quantification of order in the Lennard–Jones system. *J. Chem. Phys.* **118**:2256–2263.

87. Harn, N., C. Allan, C. Oliver, and C. R. Middaugh, 2007. Highly concentrated monoclonal antibody solutions: direct analysis of physical structure and thermal stability. *J. Pharm. Sci.* **96**:532–546.

88. Tomicki, P., R. L. Jackman, and D. W. Stanley, 1996. Thermal stability of metmyoglobin in a model system. *Lebensm.-Wiss. u.-Technol.* **29**:547–551.

89. Iametti, S., B. D. Gregori, G. Vecchio, and F. Bonomi, 1996. Modifications occur at different structural levels during the heat denaturation of β-lactoglobulin. *Eur. J. Biochem.* **237**:106–112.

90. Hamada, D., and C. M. Dobson, 2002. A kinetic study of β-lactoglobulin amyloid fibril formation promoted by urea. *Prot. Sci.* **11**:2417–2426.

91. Park, C., and R. T. Raines, 2000. Dimer formation by a "monomeric" protein. *Prot. Sci.* **9**:2026–2033.

92. Gotte, G., F. Vottariello, and M. Libonati, 2003. Thermal aggregation of ribonuclease A. *J. Biol. Chem.* **278**:10 763–10 769.

93. Libonati, M., and G. Gotte, 2004. Oligomerization of bovine ribonuclease A: Structural and functional features of its multimers. *Biochem. J.* **380**:311–327.

94. Liu, Y., and D. Eisenberg, 2002. 3D domain swapping: As domains continue to swap. *Prot. Sci.* **11**:1285–1299.

95. Ferrone, F. A., 2004. Polymerization and sickle cell disease: a molecular view. *Microcirculation* **11**:115–128.

96. Vaiana, S. M., M. A. Rotter, A. Emanuele, F. A. Ferrone, and M. B. Palma-Vittorelli, 2005. Effect of T-R conformational change on sickle-cell hemoglobin interactions and aggregation. *Proteins* **58**:426–438.

97. Johnson, C. R., and E. Freire, 1996. Structural stability of small oligomeric proteins. *Techn. Prot. Chem.* **7**:459–467.

98. Kendrick, B. S., T. Li, and B. S. Chang, 2002. In *Rational Design of Stable Protein Formulations*, pp. 61–83. New York: Plenum Press.

99. Hou, L., F. Lanni, and K. Luby-Phelps, 1990. Tracer diffusion in F-actin and ficoll mixtures: toward a model for cytoplasm. *Biophys. J.* **58**:31–43.

100. Wang, W., 1999. Instability, stabilization, and formulation of liquid protein pharmaceuticals. *Int. J. Pharmaceut.* **185**:129–188.

101. Zimmermann, S. B., and A. P. Minton, 1993. Macromolecular crowding: biochemical, biophysical and physiological consequences. *Ann. Rev. Biophys. Biomol. Struct.* **22**:27–65.

102. Minton, A. P., 1998. Molecular crowding: analysis of effects of high concentrations of inert cosolutes on biochemical equilibria and rates in terms of volume exclusion. *Meth. Enzymol.* **295**:127–149.

103. Hall, D., and A. P. Minton, 2002. Effects of inert volume-excluding macromolecules on protein fiber formation I: Equilibrium models. *Biophys. Chem.* **98**:93–104.

104. Sasahara, K., P. McPhie, and A. P. Minton, 2003. Effect of dextran on protein stability and conformation attributed to macromolecular crowding. *J. Mol. Biol.* **326**:1227–1237.

105. O'Connor, T. F., P. G. Debenedetti, and J. D. Carbeck. 2004. Simultaneous determination of structural and thermodynamic effects of carbohydrate solutes on the thermal stability of ribonuclease A. *J. Am. Chem. Soc.* **126**:11 794–11 795.

2

Observations on the mechanics of a molecular bond under force

L. B. FREUND

Introduction

In the sections that follow, two issues focusing on the mechanics of a molecular bond under the action of an applied force are addressed. This discussion is motivated by the behavior of the noncovalent molecular bonds that account for adhesion of eukaryotic animal cells to extracellular matrix or to other cells, particularly bonding of transmembrane integrins to their ligands. The integrins are large, compliant molecules and the results of experiments on their behavior are reported in the literature. The discussion itself, however, is generic within the context of physical chemistry.

The first issue considered is the transient response of a single molecular bond to an applied force. The configuration of the system is represented by one or more stochastic variables and by a deterministic variable, the latter accounting for the force input. Bonding is understood to be the confinement of the configuration of the bond within an energy well, as represented by the stochastic variables, which exists as a feature of an overall energy landscape, and the role of the force is to distort that landscape. The system functions in a thermal environment, but the time required for bond separation is long compared to the thermal relaxation time of a single step in the Brownian motion of a small particle in that environment. The bond state is characterized by a probability distribution of possible configurations over the bond energy landscape, and the evolution of that distribution is governed by the Smoluchowski partial differential equation. The system is analyzed to extract information on the most

Statistical Mechanics of Cellular Systems and Processes, ed. Muhammad H. Zaman. Published by Cambridge University Press. © Cambridge University Press 2009.

probable time for bond separation, the maximum force that can be sustained by the bond, and the sensitivity of bond separation behavior to the rate of loading and bond well features, for example.

The second issue considered is the stability of a molecular bond by which a membrane is attached to a substrate. Again, the bond is represented by an energy well in the landscape representing the dependence of bond energy on its configuration. In this case, the system is in thermal equilibrium with its environment, so the membrane position fluctuates continuously as a result. In the course of random fluctuations, the membrane exerts a fluctuating force on the bond which is equal but opposite to the constraint force acting on the membrane. The particular question posed concerns the largest edge-to-edge span that the membrane can have without overwhelming the bond through thermal fluctuations. The phenomenon of interest is a feature of steady behavior and, consequently, it is studied by means of classical statistical mechanics. The result of the analysis is of potential interest in understanding the spacing of integrins within a focal adhesion zone as it is formed in the course of adhesion of a biological cell.

The two issues addressed are fundamentally different aspects of the same physical system. Although each can be understood in a fairly general way, there are few results of a general nature that can be inferred about the transient behavior of the full system of an interacting bond and the membrane to which one of the bound molecules is attached. However, the results presented here are seen as steps in the direction of achieving that objective.

Forced separation of a molecular bond

In this section, we examine the situation of a molecular bond being acted upon by a loading apparatus that tends to pull the bond apart. The bond itself is described by means of an expression of bond energy versus a reaction coordinate, commonly known as the *energy landscape* of the bond. The loading apparatus is described by means of a deformable element linking the reaction coordinate to a point at which a time-dependent constraint is externally imposed. The force on the bond is understood to be the force needed to maintain this constraint. The reaction coordinate is stochastic and can be described only in terms of a time-dependent probability distribution whereas the imposed constraint is deterministic.

That such a process can be observed has been demonstrated by Evans *et al.* [1], who have reported measurements of the dependence of the maximum force required to overcome the bond on the rate of application of that force. An intriguing question concerns the way in which the behavior observed is related

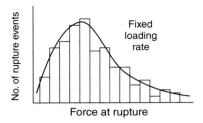

Fig. 2.1 A generic diagram illustrating the way in which data were reported by Evans and Ritchie [2] on the observed behavior of isolated molecular bonds under the action of steadily increasing force. The response of a large number of individual bonds under nominally identical loading conditions was represented as a histogram in this form.

to the properties of the bond being overcome. Equilibrium statistical mechanics is too restrictive for addressing such questions, so the process is viewed more generally as being a stochastic transport phenomenon.

The general character of the data on bond rupture strength reported by Evans and Ritchie [2] is as follows. Suppose a large number (several hundred) of rupture events are observed for nominally identical bond pairs and nominally identical rates of loading up to rupture. If the results are presented in the form of a histogram of number of rupture events that occurred within each of numerous finite increments of applied force, the results take the form illustrated in Fig. 2.1. The spread of the data along the force axis suggests that the process is statistical in nature. The main influence of an increase in the rate of loading was found to be a stretch of the distribution in the direction along the force axis. In particular, the peak in the distribution moves to a larger force value as the rate of loading increases. The purpose in this section is to discuss a theoretical framework for interpreting such observations.

A model of bond breaking

In selecting features to be incorporated into a model of the bond rupture process, we follow the lead of Evans and Ritchie [2] in several respects. The bond itself is idealized as a one-dimensional energy landscape, as depicted in Fig. 2.2. The abscissa x in this diagram is the so-called reaction coordinate, a variable that determines the configuration of the bond. While the process of bond rupture might require extensive conformational changes of the molecules, such as twisting or unfolding, the energy input that effects these changes is assumed to be delivered by a force working through the translational coordinate x.

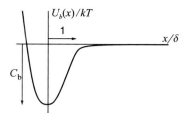

Fig. 2.2 From the energy point of view, adhesion is represented as an energy well that is accessible in some configurations of the bond pair. The diagram illustrates the general features of such a well that are incorporated in the model. The bond energy U, which is a function of configuration as represented by a reaction coordinate x, is described mainly in terms of the depth $C_b kT$ and breadth δ of the well. The diagram shows the system landscape in the vicinity of the well in terms of normalized parameters.

The quantity $U_b(x)$ in Fig. 2.2 represents the interaction energy of the bond pair at reaction coordinate x, and this energy is normalized by the thermal unit kT where k is the Boltzmann constant and T is the absolute temperature. The reaction coordinate is normalized by the half-width δ of the well. The depth of the well C_b is the reduction in free energy of the system upon bond formation at low temperature; typically $C_b \sim 10$. When the molecules are well separated, the interaction energy is zero; the molecules are fully separated at values of the reaction coordinate greater than δ.

The loading apparatus is understood to be the link between the stochastic domain of the molecule and the deterministic constraint being enforced in order to separate the molecules. It is incorporated into the model by means of a spring, either linear or nonlinear, to represent the potentially significant feature of loading mechanism compliance. As illustrated in the sketch in Fig. 2.3, the position of one end of the loading apparatus is the reaction coordinate x and the position $y(t)$ of the other end is prescribed. Both x and y are measured with respect to their initial locations before the force is applied. The applied force that produces the rupture of the bonds is more appropriately considered as a quantity derived through its effect rather than as a fundamental variable in a thermodynamic framework and, as such, it does not appear explicitly in the model. Instead, we determine this force as that action which produces the energy changes in the system in the course of its response to the imposed motion $y(t)$.

The energy landscape of the system is represented by a surface over the plane spanned by the reaction coordinate x and the loading coordinate y. It is the sum of two contributions, one of which represents bond behavior as depicted in Fig. 2.2 and the other of which represents the properties of the

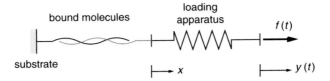

Fig. 2.3 The diagram is a schematic of the entire system. The configuration of the bond is represented by a single time-dependent reaction coordinate x, which varies stochastically, and a time-dependent coordinate y which is specified a priori. The essential properties of the bond are C_b and δ, and the properties of the loading apparatus are also specified. The applied force $f(t)$ is calculated to be the force necessary to maintain the specified motion $y(t)$.

loading apparatus. As an illustration, suppose that the former is described by

$$\frac{U_b(x)}{kT} = u_b(\xi) = \begin{cases} -C_b + 2C_b\xi^2, & \xi < 0.5 \\ -2C_b(\xi - 1)^2, & 0.5 \le \xi \le 1 \\ 0, & 1 < \xi \end{cases} \tag{2.1}$$

where $\xi = x/\delta$ is the nondimensional variable introduced in Fig. 2.2 and C_b is the depth of the well. The latter is a linear spring described by

$$\frac{U_s(x, y)}{kT} = u_s(\xi, \eta) = \frac{1}{2}\kappa(\eta - \xi)^2 \tag{2.2}$$

where $\eta = y/\delta$ and the nondimensional parameter κ is the spring stiffness normalized by kT/δ^2. The energy landscape is then the surface $U(x, y) = U_b(x) + U_s(x, y)$ over the x, y-plane or, in normalized parameters, $u(\xi, \eta) = u_b(\xi) + u_s(\xi, \eta)$ over the ξ, η-plane. The features of this surface can be visualized in several ways. Here, we do so by plotting cross-sections of the surface for several values of η. This is illustrated in Fig. 2.4 for $C_b = 10$ and the stiffness value $\kappa = 1$. The entire surface near $\xi = 0$ translates in the direction of the energy axis (that is, "upward" in the figure) as η increases, so the quantity plotted is actually $u(\xi, \eta) - \frac{1}{2}\kappa\eta^2$ to compensate for this translation. In their original report, Evans and Ritchie [2] assumed that the external force was applied directly to the reaction coordinate, whereas a spring to represent loading compliance was adopted in the qualitative discussion of force spectroscopy given by Evans and Calderwood [3].

The main features of the evolving surface are evident in Fig. 2.4. The surface has a zero energy trough along the line $y = x$. This is the state with the spring relaxed and the molecules fully separated. This is where the system ends up on the energy surface after complete bond separation.

Initially, with $y = 0$, the bond well is the expected location or state of the system. As y increases, the well is elevated to higher energy levels relative to the

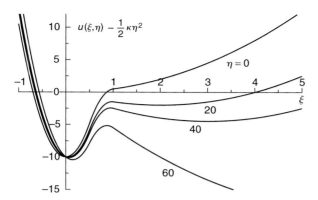

Fig. 2.4 The two-dimensional energy landscape of the system $u(\xi, \eta)$ is illustrated by means of several cross-sections with $\eta =$ constant in terms of nondimensional coordinates $\xi = x/\delta$ and $\eta = y/\delta$. The level of the landscape rises quite rapidly with increasing η because energy is continually added by the applied force. Consequently, a time-dependent translation is subtracted for each section so as to represent all sections in the same coordinate system.

zero energy trough, and more so at the left end than at the right end, thereby giving the impression that the local landscape is rotated in a clockwise sense. This distortion promotes the flux of states from within the well over the barrier and in the direction of the zero energy trough. This is more or less the same viewpoint introduced by Evans and Ritchie [2], except that they had the applied force working directly through the reaction coordinate. Some differences will emerge in the analysis of the model below, however.

It should be recognized that this representation of the energy landscape of the molecular bond in terms of one or two degrees of freedom is extremely coarse, in light of the structural complexity of the molecules whose interactions form the bond. However, at this point in the development of the topic, there is little basis for choosing a more elaborate representation, although numerical studies of molecular interactions such as those reported in [4] should begin to provide guidance for doing so in the future. In an interesting discussion of breaking of so-called catch bonds between certain types of molecules due to force, Thomas [5] has suggested that such structures may account for the unusual behavior exhibited by these molecular pairs.

With this picture of the energy landscape in place, the question to be addressed is the following. If the state of the system initially resides in the bond interaction well when $y = 0$ and if the coordinate y is specified to be some increasing function of time thereafter, what is the time history of the force acting on the bond and of the probability distribution of bond configurations?

The state of the bond

In view of the stochastic nature of bond behavior, the state of the bond is represented by a time-dependent probability distribution which is a function of the reaction coordinate x, say $\rho(x, t)$. If a large number of identical systems are observed under nominally identical constraints, the corresponding distribution in values of the reaction coordinate at a certain instant of time provides a *density of states* for the system at that time. If this density of states is normalized so as to represent the fractional distribution of states, it can be interpreted as a *probability distribution* for the current value of reaction coordinate in a particular system. In the discussion that follows, the terms density of states and probability distribution are used interchangeably with the understanding that they refer to one and the same thing. To ensure that all states are taken into account over time, this probability distribution $\rho(x, t)$ on the finite interval $x_0 < x < x_1$ is constrained by the conservation condition

$$\int_{x_0}^{x_1} \rho(x, t)\, dx - \int_0^t j_0(s)\, ds + \int_0^t j_1(s)\, ds = 1 \tag{2.3}$$

identically in time, where j_0 and j_1 represent fluxes of states inward at $x = x_0$ and outward at $x = x_1$.

In the thermal environment of a solvent, the molecules are continually bombarded by impulses, each of which delivers an energy on the order of kT. This energy is dissipated back into the environment through the viscous resistance of the solvent in the course of motion of the molecules over a small distance in a short time. If the depth of the interaction energy well in the energy landscape is significantly greater than kT, then the likelihood of escape from the well (that is, debonding) on the timescale of individual excitations is very small and the probability distribution evolves diffusively, rather than ballistically. In other words, transition of a configuration out of the well is an improbable event. In this setting, the diffusion coefficient D is expressible in terms of the viscosity γ of the solvent and the thermal energy unit kT through the Einstein relation $D = kT/\gamma$ [6]. This is the point of view of chemical kinetics that was developed through the work of Smoluchowski and Kramers, among others, as described in systematic detail by Risken [7] and Hanngi [8]. This viewpoint provides the basis for extracting a partial differential equation governing the probability distribution.

The local flux of states in the direction of increasing x, here denoted by $j(x, t)$, has the form

$$j(x, t) = -\frac{kT}{\gamma}\left[\frac{\partial \rho}{\partial x}(x, t) + \rho(x, t)\frac{\partial u}{\partial x}(x, t)\right] \tag{2.4}$$

where $u(x, t) = U(x, y(t))/kT$. The first term represents the diffusive transport of states in a force-free environment due to a thermally activated random walk whereas the second term represents directed transport in the downhill direction on the local energy landscape. The requirement of local conservation of states,

$$\frac{\partial \rho}{\partial t}(x,t) + \frac{\partial j}{\partial x}(x,t) = 0, \tag{2.5}$$

is equivalent to (2.3). Combining (2.4) with the conservation condition leads to the Smoluchowski partial differential equation [7] governing the probability density,

$$\frac{\partial \rho}{\partial t}(x,t) = \frac{kT}{\gamma} \frac{\partial}{\partial x} \left[\frac{\partial \rho}{\partial x}(x,t) + \rho(x,t) \frac{\partial u(x,t)}{\partial x} \right]. \tag{2.6}$$

This equation must be augmented by an initial condition and suitable boundary conditions in order to obtain a solution. Boundary conditions will be specified in connection with a particular case of interest.

As an initial condition at time $t = 0$, we assume that the distribution of states is confined to the bond well and that the flux of states is zero. This implies that the initial condition is determined by setting the quantity enclosed in square brackets on the right side of (2.6) equal to zero. The initial probability distribution is then determined as the solution of the ordinary differential equation $\partial_x \rho(x,0) + \rho(x,0)\partial_x u(x,0) = 0$ and the normality condition (2.3) to be

$$\rho(x,0) = \frac{e^{-u(x,0)}}{\int_{\mathcal{L}} e^{-u(x',0)} \, dx'} \tag{2.7}$$

where $\mathcal{L}$ denotes the spatial range of the reaction coordinate over the landscape.

We again denote the half-width of the bond well by the length δ and normalize all distances by this characteristic dimension. Suppose we also normalize time by the diffusion time over the distance δ, which is $\gamma \delta^2 / kT$. If a nondimensional reaction coordinate ξ and a nondimensional time τ, as defined by

$$\xi = x/\delta \quad \tau = tkT/\gamma \delta^2 \tag{2.8}$$

are introduced, then the Smoluchowski equation takes on the parameter-free form

$$\frac{\partial \rho}{\partial \tau}(\xi,\tau) = \frac{\partial}{\partial \xi} \left[\frac{\partial \rho}{\partial \xi}(\xi,\tau) + \rho(\xi,\tau) \frac{\partial u}{\partial \xi}(\xi,\tau) \right]. \tag{2.9}$$

This equation is to be enforced over the spatial range $\mathcal{L}$ of the landscape, which is $-1 < \xi < 3$ in Fig. 2.4, and over all time $\tau > 0$. When augmented by the initial condition (2.7) and suitable boundary conditions, this partial differential equation can be solved approximately by any number of numerical strategies. However, before proceeding in that direction it is worthwhile to examine the range of the dimensionless independent variables that are relevant to the phenomenon being considered.

The important range for ξ is set by the range of the physical coordinate x required to span the interaction energy well and its surroundings, which is a distance equal to several times the length δ. Consequently, the range of ξ of interest is roughly $-1 < \xi < 3$. Similarly, the important range for τ is set by the elapsed time from the instant a force is first applied to the time of bond separation, which is typically on the order of 10 ms or greater. The corresponding range of τ can be estimated from (2.8). Assuming the factor kT/γ to have a value equal to the diffusion coefficient for water, or roughly $10^{-10}\,\mathrm{m^2/s}$, and δ to be about 5 nm, then the range of τ corresponding to elapsed time of 10 ms is roughly 10^2. This implies that the probability distribution $\rho(\xi, \tau)$ varies slowly on the landscape in the τ direction compared to its variation in the ξ direction for comparable increments in ξ and τ. This observation suggests, in turn, that the rate of change of ρ on the left side of (2.9) is very small compared to the other contributions. It is assumed to be negligibly small for present purposes so that the governing equation reduces to

$$\frac{\partial}{\partial \xi}\left[\frac{\partial \rho}{\partial \xi}(\xi, \tau) + \rho(\xi, \tau)\frac{\partial u}{\partial \xi}(\xi, \tau)\right] \approx 0 \tag{2.10}$$

with τ viewed as a parameter. In this way, the partial differential equation has been reduced to an ordinary differential equation to be solved for all values of time τ.

A formulation similar to the present development was proposed by Heymann and Grubmuller [9] but with one significant difference. No specific landscape profile was presumed at the outset. Instead, conclusions were drawn by introducing reaction rate coefficients early in the analysis and, subsequently, the shape of the landscape profile was extracted from observational data by means of the results obtained.

The probability distribution $\rho(\xi, \tau)$

In an important contribution to physical chemistry, Kramers (see Ref. [7]) showed how the "on" and "off" rate coefficients of elementary rate reaction theory can be determined from any presumed time-independent energy landscape for a reaction occurring under diffusion-dominated circumstances. Through this result, it became possible to determine the influence of distance between an equilibrium state and the transition state in terms of reaction coordinate, the influence of attempt frequency on rate, the influence of curvature of the energy surface at the transition state, and so on. As introduced originally in elementary reaction rate theory, the coefficients depend on the energies at the equilibrium states and the transition state, but not on the shape of the landscape between these energy levels. By superimposing a constant uniform flux of probability density from one equilibrium state to another, the governing Smoluchowski partial differential equation could be reduced to an ordinary differential equation. The reduction can be rendered exact in the case of the *time-independent* landscape through an insightful selection of boundary conditions, and this uniform flux was shown to be exactly the "on" or "off" rate coefficient. The reasoning is similar here, but the time dependence of the interaction energy landscape and the fact that probability density is not conserved within the interval of interest necessitate adoption of a different point of view.

The first integral of the ordinary differential equation (2.10) is

$$\frac{d\rho}{d\xi}(\xi,\tau) + \rho(\xi,\tau)\frac{du}{d\xi}(\xi,\tau) = -j(\tau) \tag{2.11}$$

with parameter of integration $j(\tau)$. While this equation must be satisfied throughout $\xi_0 < \xi < \xi_1$, it is not essential for each term to be continuous. We expect the probability density $\rho(\xi,\tau)$ and the gradient in the landscape $\partial_\xi u(\xi,\tau)$ to be continuous on physical grounds. On the other hand, it is possible that $j(\tau)$ is only piecewise constant, in which case its discontinuities must be exactly offset by corresponding discontinuities in $d\rho/d\xi$. That this possibility exists is not a trivial observation. If $j(\tau)$ is required to be a spatially *uniform* flux throughout the entire interval and if we enforce the boundary condition that the flux at $\xi = \xi_0$ is equal to zero, then $j(\tau)$ must be equal to zero throughout the interval and no solution with nonzero flux can be found. The ordinary differential equation is readily solved within any interval for which $j(\tau)$ is a constant and we proceed accordingly.

Several specific points in the range of ξ that covers the landscape have particular significance, and these are identified in Fig. 2.5. The point $\xi = \xi_0$ locates the extreme left end of the range of interest; this is $\xi_0 = -1$ in the present instance. The point $\xi = \xi_a(\tau)$ identifies the position of the local minimum in the interaction energy well profile; the location of this minimum is time dependent, in general. The point $\xi = \xi_c(\tau)$ identifies the position of the local maximum in the landscape profile, that is, the transition state; this location is also time dependent. Finally $\xi = \xi_1$ locates the extreme right end of the range of interest; for the profile being considered here, $\xi_1 = 3$ although the choice is arbitrary to a certain degree within $\xi_c(\tau) < \xi$.

There are some restrictions on the definitions of the points $\xi_a(\tau)$ and $\xi_c(\tau)$ for certain ranges of system parameters. For example, in very early times, a local maximum in the energy landscape may not exist for some specified $\eta(\tau)$. Likewise, at late times, the local minimum and maximum points may coalesce in some cases. These circumstances are unusual and neither is a factor in the example considered below.

We could proceed to formally solve the differential equation (2.11). However, it is more enlightening to construct a solution by superposition after noting the physical roles played by the homogeneous solution and the particular solution.

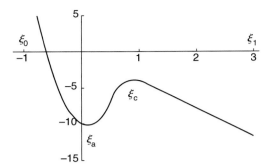

Fig. 2.5 The graph illustrates a constant time cross-section of the energy landscape of the system in normalized coordinates, identifying several values of spatial coordinate ξ that identify particular points of significance in the analysis. The point ξ_0 is the extreme left end of the range of interest; the flux of states is zero at this point. The point ξ_1 is the extreme right end of the interval of interest; the density of states is zero at this point. The points ξ_a and ξ_c are local minimum and maximum energies in this section, respectively; the locations of these points vary with time.

The quantity of primary interest is the probability that the state of the bond remains in the well. Initially, at $\tau = 0$, this probability is one, and we expect it to decrease as τ increases. Let this probability be represented by $\rho_a(\tau)$, that is,

$$\rho_a(\tau) = \int_{\xi_0}^{\xi_c(\tau)} \rho(\xi, \tau)\, d\xi, \tag{2.12}$$

where the subscript indicates that the probability is connected with the energy well at $\xi = \xi_a(\tau)$. The integrating factor for this differential equation is $e^{u(\xi,\tau)}$, and it is convenient to introduce a compact notation for representing definite integrals of this factor and its inverse. For this purpose, we introduce

$$I_{\pm}(\xi_-, \xi_+) = \int_{\xi_-}^{\xi_+} e^{\pm u(\xi,\tau)}\, d\tau, \quad \xi_+ \geq \xi_-. \tag{2.13}$$

The value is identically zero if $\xi_+ \leq \xi_-$. We see that a distribution of states within the well defined by

$$\rho(\xi, \tau)\Big|_{\text{homog}} = \rho_a(\tau)\frac{e^{-u(\xi,\tau)}}{I_-(\xi_0, \xi_{c+})} \tag{2.14}$$

provides a homogeneous solution of the differential equation over $\xi_0 < \xi < \xi_{c+}$, having the essential attributes we have noted. It is clear that the homogeneous solution generates no flux of probability density.

The role of the particular solution is to drain those states represented by (2.14) out of the well centered at ξ_a and to pass them over the barrier at ξ_c. It was noted above that a particular solution with *uniform* flux throughout the range $\xi_0 < \xi < \xi_c$ is out of the question. Therefore, we adopt a particular solution with *zero* flux within $\xi_0 < \xi < \xi_a$ and with the *uniform* flux $j(\tau)$ between $\xi = \xi_a$ and some point beyond the transition state at $\xi = \xi_c$, say $\xi = \xi_{c+}$. Consequently, any choice of particular solution within $\xi_a < \xi < \xi_{c+}$ must satisfy $\rho(\xi_a, \tau)|_{\text{part}} = 0$ to render the probability density continuous. A distribution satisfying all the requirements is

$$\rho(\xi, \tau)\Big|_{\text{part}} = -j(\tau)e^{-u(\xi, \tau)} \int_{\xi_a}^{\xi} e^{u(\xi', \tau)} \, d\xi' = -j(\tau)e^{-u(\xi, \tau)}I_+(\xi_a, \xi) \tag{2.15}$$

so that the complete solution within $\xi_a < \xi < \xi_{c+}$ is

$$\rho(\xi, \tau) = e^{-u(\xi, \tau)}\left[\frac{\rho_a(\tau)}{I_-(\xi_0, \xi_c)} - j(\tau)I_+(\xi_a, \xi) \right]. \tag{2.16}$$

Then, the boundary condition $\rho(\xi_{c+}, \tau) = 0$ requires that

$$j(\tau) = \frac{\rho_a(\tau)}{I_-(\xi_0, \xi_c)I_+(\xi_a, \xi_{c+})} \equiv \rho_a(\tau)g(\tau). \tag{2.17}$$

In obtaining the numerical results reported in the section which follows, the value of ξ_{c+} was chosen to be 1.5.

It is important to recognize the physical significance of the flux $j(\tau)$. In obtaining (2.17), it was understood that this flux exists over $\xi_a < \xi < \xi_{c+}$. At the point ξ_{c+}, the flux represents the rate at which states within a large ensemble of identical systems are flowing out of their energy wells; an understanding of this rate is one of the main goals of the model. At the point ξ_a, on the other hand, no compensating external source of states exists, as it does in the Kramers model. Instead, the flux draws states from the distribution of states (2.14) *within the well*.

With this interpretation established, we can readily determine the time history of $\rho_a(\tau)$. This quantity represents the probability that the state of the bond is still in the well. Considering its initial value $\rho_a(0) = 1$ and the flux $j(\tau)$ of states out of the well together, the conservation condition (2.3) implies

$$j(\tau) = -\frac{d\rho_a}{d\tau}(\tau) = -\frac{d}{d\tau}\int_{\xi_0}^{\xi_c} \rho(\xi, \tau) \, d\xi. \tag{2.18}$$

Even though the full conservation equation (2.9) was abandoned on the basis of the slow variation of the density distribution field with τ, it is still necessary to ensure that states are conserved over time. This is done by (2.18). Together, (2.17) and (2.18) yield an ordinary differential equation for $\rho_a(\tau)$ which, with the initial condition $\rho_a(\tau) = 1$, can be solved by inspection to find the result expressed in terms of a definite integral of the function $g(\tau)$ defined in (2.17) as

$$\rho_a(\tau) = \exp\left[-\int_0^{\tau} g(s) \, ds\right], \quad 0 < \tau < \infty. \tag{2.19}$$

In this way, the history of the probability that the bond is within the well is expressed in terms of the energy landscape, loading rate and loading apparatus compliance in the form of this result.

We recognize that $g(\tau)$ defined in (2.17) is the quantity commonly identified as the *off-rate* for chemical states within the well, according to elementary rate reaction theory. (In dimensional terms, $g(\tau)kT/\gamma\delta^2$ is the off-rate.) The integral defining the off-rate is a relative measure of the details in the energy landscape between the well minimum at $\xi_a(\tau)$ and the point $\xi = \xi_{c+}$ beyond the barrier maximum at $\xi_c(\tau)$. The result (2.19) therefore provides an expression for the off-rate of the separation process in terms of the evolving shape of the landscape, thereby generalizing the corresponding result of Kramers to the case of a time-dependent landscape. The integral appearing in (2.19) is well suited to approximate evaluation by means of the Laplace asymptotic method. A systematic study of the dependence of the off-rate on applied force, loading rate, loading stiffness, and the features of the energy landscape could be enlightening, but that task is beyond the scope of the present discussion.

Implications

In the above illustration of an energy landscape, the bond itself was described by (2.1) and the loading apparatus by a linear spring with normalized stiffness κ. The applied force is that force which causes the loading point to move with the prescribed displacement $\eta(\tau)$. For purposes of illustration, we now assume that

$$\eta(\tau) = \beta\tau \tag{2.20}$$

where β is the constant nondimensional speed at which the load point moves. In this case, evaluation of the expression for $\rho_a(\tau)$ is simplified significantly by recognizing that $g(\tau)$ depends on β and τ only through the product $\beta\tau$.

In order to demonstrate the nature of (2.19), the result of numerical evaluation of the integral defining $\rho_a(\tau)$ for the particular case with $\kappa = 0.5$ and $\beta = 0.1$ is illustrated in Fig. 2.6 by the solid line. As expected, the probability that the bond remains intact is near one for some length of time, and then undergoes a transition to very small values within a relatively short period of time.

To get a sense of the quality of the approximation represented by the result (2.19), we analyzed the original partial differential equation (2.9) numerically in the interval $\xi_0 < \xi < \xi_1$, with the boundary conditions $j(\xi_0, \tau) = 0$ and $\rho(\xi_1, \tau) = 0$ and the initial condition (2.7) (in normalized form). The finite element method was adopted as the basis for the calculation. For purposes of comparison with $\rho_a(\tau)$, the probability that the bond state was in the well at any time τ was

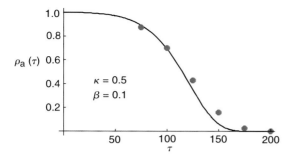

Fig. 2.6 The solid curve shows the time history of probability that the state of the system is still in the well, implying the bond is fast for a particular set of system parameters. The time of bond rupture is chosen to be the time at which the probability is equal to one half. The discrete points are obtained by numerical solution of the full Smoluchowski partial differential equation for the same set of system parameters, with the agreement between the two results illustrating the quality of the approximate solution.

determined from the solution by evaluating the integral

$$\rho_a(\tau)\Big|_{\text{fem}} = \int_{\xi_0}^{\xi_c(\tau)} \rho(\xi, \tau)\, d\xi \tag{2.21}$$

with the density $\rho(\xi, \tau)$ obtained through numerical solution. The results for the case when $\kappa = 0.5$ and $\beta = 0.1$ are shown by the discrete points in Fig. 2.6. As is evident from the figure, the results obtained by numerical solution of the full Smoluchowski equation and from evaluation of the explicit approximation (2.19) are in good agreement for the particular case illustrated.

Suppose that we adopt the normalized time τ_* defined by

$$\rho_a(\tau_*) = \tfrac{1}{2} \tag{2.22}$$

as the most likely time for bond separation. This is the elapsed time when the fractional number of states that remain within the well has fallen to one half.

Under these circumstances, (2.22) is an expression for the most probable time τ_* of bond separation in terms of loading rate β and loading stiffness κ, all in normalized form. In order to illustrate the response anticipated by this model, representative features of this relationship have been extracted and these are shown in Fig. 2.7. The figure shows the dependence of the natural log of the most probable time for bond separation on the natural log of the loading rate for five values of loading stiffness. Each curve in the graph consists of straight line segments joining pairs of successive data points.

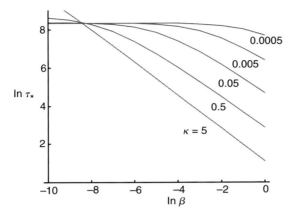

Fig. 2.7 The plot illustrates the dependence of the time τ_* required to rupture the bond on the rate of loading β for three values of loading apparatus stiffness κ, all quantities being normalized as described in the text.

Figure 2.7 reveals wide variability in response to force, depending on the loading rate and the stiffness of the loading apparatus. For extremely stiff loading (very large κ), we can expect that the value of τ_* will be determined simply from the relationship $\tau_*\beta = 1$, that is, from the condition that the bond will separate when the load point has moved a distance equal to the half-width of the bond well. Such behavior would appear in the graphs as a straight line with slope -1 that passes through the origin. The plot for $\kappa = 5$ has features that are close to this behavior. The important role of the loading compliance has been emphasized in several recent studies of bond rupture under force, as in Ref. [10] for example.

At the opposite extreme, for an extremely soft loading (very small κ), we expect that the value of τ_* will be equal to the dissociation time for the bond in the absence of applied force. Such a result would appear in the graph as a horizontal line with position defined by the dissociation time. The plot for $\kappa = 0.0005$ exhibits features that imply this behavior. The curve has some nonzero curvature for loading rates at the higher end of the range covered in the figure. For a value of loading rate less than about e^{-2}, however, the curve is virtually flat at a normalized time of approximately e^8. Therefore, this time can be interpreted as the dissociation time for the bond.

The curves in the figure for values of stiffness between $\kappa = 5$ and $\kappa = 0.0005$ reflect a gradual transition in behavior between the extremes identified for very large and very small loading stiffness. It is not surprising that the curves for adjacent values of κ should intersect at some point corresponding to a low rate of loading. The reason is simply that, for very low rates of loading, the influence of loading stiffness is to *restrict* the fluctuations in reaction coordinate for the bond. As a result, the bond life at extremely low rate will be *greater than* the natural dissociation rate. Therefore, it is not unreasonable to expect the time at the crossover point to correspond to the bond dissociation time. If this is so for each successive pair of curves, it follows that the whole family of curves necessarily shares a common crossover point, as illustrated in Fig. 2.7.

The force acting on the bond is the force transmitted to it by the spring; for a linear spring, this is the stiffness times the spring extension. In the present case, the expected value of the nondimensional force, say $\phi(\tau)$, acting on the bond at time τ is

$$\phi(\tau) = \kappa[\eta(\tau) - \xi_a(\tau)] = \frac{\kappa \beta \tau}{1 + \kappa/40} \tag{2.23}$$

in terms of nondimensional quantities.

One way to represent the influence of applied force on bond separation is suggested by Fig. 2.1. The ordinate in the graph (which is only a sketch of the type of experimental data reported by Evans and Ritchie [2]) is the number of times a bond separated at an applied force level within a certain range. If the ordinate is interpreted instead as the *fraction of total events* which resulted in separation of the bond within each force range, then the continuous approximation to the histogram is equivalent to $-d\rho_a/d\phi$ versus ϕ in the present notation. Both $\rho_a(\tau)$ and $\phi(\tau)$ as functions of time are established by the solution for any values of the system parameters. Consequently, the same kind of graphs can be extracted from the model for any values of β and κ by plotting $-\rho_a'(\tau)/\phi'(\tau)$ versus $\phi(\tau)$ parametrically in τ. Essentially the same way of representing this result was applied in Ref. [11] where bond breaking by means of the atomic force microscope was analyzed for unfolding of ubiquitin.

The bond response to loading is illustrated in this way in Fig. 2.8 for loading stiffness $\kappa = 0.5$ and for four values of loading rate, $\beta = 0.001, 0.001, 0.1$, and 1. The similarities of the result to the behavior illustrated in the sketch in Fig. 2.1 are evident.

Perhaps the most significant aspect of the results in Fig. 2.8 concern the dependence of the maximum force required to overcome a bond on the rate at which that force is applied. The force is maximum when $-\rho_a'(\tau)$ is a maximum, which occurs at the time $\tau = \tau_*$. Consequently, the maximum force will be denoted by ϕ_*. For the value of loading rate $\beta = 1$, the maximum force is $\phi_* \approx 9$ whereas for $\beta = 0.01$ the value is $\phi_* \approx 3$.

In an analysis of bond rupture due to increasing force, Evans [12] concluded that the maximum force would vary with loading rate β as $\ln \beta$. We can check to see how closely the prediction of the present model follows that expectation by extracting it from Fig. 2.8. Alternatively, it can be calculated directly by using

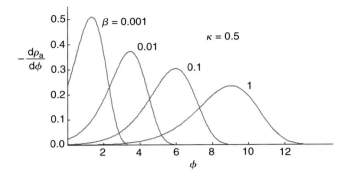

Fig. 2.8 The graphs illustrate the sensitivity of the probability that the bond remains secure to the force applied to the bond for stiffness $\kappa = 0.5$ and for four values of the loading rate parameter β as shown, analogous to the histogram shown in Fig. 2.1.

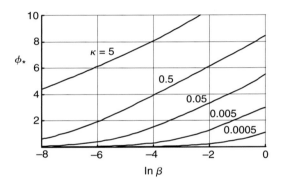

Fig. 2.9 The graphs illustrate the dependence of the critical force φ_* on $\ln \beta$ for five values of loading stiffness κ. From the plots, it is evident that the expected value of separation force is proportional to $\ln \beta$ for each value of loading stiffness, provided that the loading rate is sufficiently large.

the results for τ_* versus β from Fig. 2.7 in the expression for force in (2.23). The result is shown in Fig. 2.9. As is evident, for each value of κ, the model shows behavior that is proportional to $\ln \beta$ provided that the loading rate is above some value that depends on the stiffness. Below this value, the dependence on β is weaker than $\ln \beta$.

Stability of bonds constraining a membrane

Consider a nominally flat, square membrane with size $\Lambda \times \Lambda$ which is immersed in a thermal environment and which is positioned near a substrate

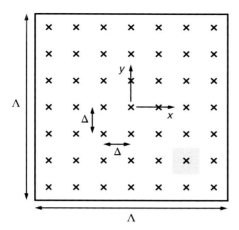

Fig. 2.10 A nominally flat portion of a flexible membrane with a regular array of potential bond sites located in a square $\Delta \times \Delta$ pattern. The behavior of the shaded square patch, which includes only a single bond, is modeled here.

with potential binding sites. The transverse deflection of the membrane $h(x, y)$ from the flat configuration depends on two independent variables x and y. The potential binding sites are indicated by the $\times$-symbols on the membrane shown in Fig. 2.10. As is evident from the diagram, these sites are arranged in a square pattern with spacing $\Delta \ll \Lambda$ in both the x and y directions of a rectangular coordinate frame in the plane of the undeformed membrane.

With reference to the configuration in Fig. 2.10, the competition between thermal fluctuation and bonding can be examined at two scales of observation. For the present discussion, we will focus on a typical square unit cell of dimension $\Delta \times \Delta$ (shown shaded in the figure) with a potential binding site at its center. This is an extreme case of the phenomenon of interest. At another level of observation, Λ is sufficiently large so that the bonds between the membrane and the substrate appear to be continuously distributed rather than discrete, a situation which has been considered recently by Lin and Freund [13]. Here, we focus on the case of a single discrete bond site.

Model problem with a single bond

A square $\Delta \times \Delta$ patch of membrane in its undeformed configuration is illustrated in Fig. 2.11. A rectangular coordinate system is introduced in the plane of the membrane with its origin at the geometrical center of the square. The local deflection of the membrane from this flat configuration at point x, y is represented by $h(x, y)$.

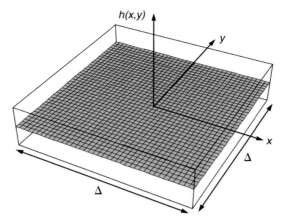

Fig. 2.11 The portion of the membrane analyzed shown in its zero-deflection reference configuration. Any admissible deflection distribution has zero average displacement and satisfies symmetry boundary conditions on the edges.

The set of admissible deflections is subjected to several constraints suggested by the physics of the configuration. For one thing, this patch is presumed to be part of a relatively large membrane. This restricts the mean transverse deflection and, for definiteness, we assume that the mean deflection is zero. In addition, this patch abuts an identical patch on each of its four sides. Expecting the behavior of each of these patches to be identical to that of every other, we assume *symmetry* boundary conditions at each of the four boundaries of the square patch. We note that *periodic* boundary conditions tend to promote anti-symmetric deformation modes; such modes have zero deflection at the bond point and, consequently, are not involved in the phenomenon of interest.

The center point of the square membrane patch provides a site for possible adhesion to the substrate. This is intended to simulate the situation of an adhesion receptor molecule at the center of the membrane and a commensurate ligand attached to the substrate below the center of the membrane. The point of view here is that, in order to be adhered at the center point, it is necessary to have that point on the membrane confined within an energy well. From the energy point of view, it is more favorable for the system to have that point within the energy well than to have it located elsewhere. On the other hand, the membrane is subject to continuous agitation as it maintains its thermal equilibrium with the large fluid heat bath in which it is immersed. This gives rise to transverse fluctuations of the membrane. Indeed, it is precisely the competition between these fluctuations, which tend to move the center point of the membrane over some relatively wide range in a random manner, and the binding interaction energy, which tends to constrain this motion to a relatively narrow range, that is of interest. This competition will be considered within the context of equilibrium thermodynamics.

Membrane fluctuations and bending resistance

Spatially nonuniform deflection from the flat configuration of the membrane involves local bending. This bending is assumed to be elastic and reversible, and this provides the mechanism of energy storage in the membrane. The in-plane behavior of the membrane is fluid-like in the sense that the material cannot support a shear stress in the xy-plane. It can support an in-plane mean normal stress, however, and the associated mean normal in-plane strain accounts for the stored elastic energy. This strain is zero in the midplane of the membrane (in the absence of membrane tension) and it increases in magnitude in proportion to the mean curvature $\partial_{xx}h + \partial_{yy}h$ with increasing normal distance from the neutral plane. These considerations give rise to an elastic bending energy of the form [14, 15]

$$U_e = \frac{1}{2}C_e \int_{-\frac{\Delta}{2}}^{\frac{\Delta}{2}} \int_{-\frac{\Delta}{2}}^{\frac{\Delta}{2}} (\partial_{xx}h + \partial_{yy}h)^2 \, dx \, dy \qquad (2.24)$$

where the bending modulus C_e has physical dimensions of force × length. In terms of the thermal energy unit kT, this modulus typically has a value in the range $20-40\,kT$ [16].

A modal approach is adopted to represent the fluctuations of the membrane, where the coefficients of the modes are the random variables governed by Boltzmann statistics. When adopting a modal approach, the mathematical question of completeness of the representation should be considered. Ideally, any modal representation of membrane deflection ought to be complete, that is, capable of accurately representing any admissible deflection distribution. From a practical point of view, however, this can impose such a heavy burden that the analysis of the model is rendered intractable. Instead, we attempt to eliminate many of the admissible modes on the basis of physical arguments in the hope of making the analysis transparent. In the present case, two such considerations are helpful. Suppose that we adopt

$$h(x, y) = \Delta \sum_{i=1}^{n_m} \sum_{j=1}^{n_m} a_{ij} \cos \frac{2\pi i x}{\Delta} \cos \frac{2\pi j y}{\Delta} \qquad (2.25)$$

as the modal representation of membrane fluctuations, where n_m^2 is the number of modes to be considered and the matrix a_{ij} represents the set of dimensionless random variables that describes the membrane shape. Two obvious features of the representation are that the number of modes is finite if n_m is finite and each of the included modes is symmetric in both x and y. The reasons underlying this selection are briefly discussed next.

Customarily, representation of the deflection of a continuous system by means of Fourier modes requires an infinite number of modes for completeness. The expected value of energy for each mode is on the order of kT according to the equipartition theorem and this observation leads to the common paradoxical situation of a system with an infinite number of modes having unbounded mean energy. However, for the expected value of energy in the ijth mode, the variance of the amplitude is found to depend on mode indices i and j according to

$$\langle a_{ij}^2 \rangle \sim \frac{kT}{2\pi^4 C_e} \frac{1}{(i^2 + j^2)^2}. \tag{2.26}$$

This observation implies that the expected value of mode amplitude is a rapidly diminishing function of increasing mode number. Because adhesion amounts to constraining mode amplitudes, this observation makes clear that the influence of modes in resisting adhesion diminishes significantly as mode number increases. More specifically, the energy interaction well that characterizes bonding potential has a characteristic width (to be represented by the symbol δ below) corresponding to compliance of the bond. Any mode for which the expected value of deflection amplitude is smaller than some fraction of the width of the well is *irrelevant* to the bonding process. Consequently, the higher modes need not be taken into account, and n_m can be chosen to be a relatively small number without risk of overlooking important aspects of response.

As has already been noted, the representation (2.25) includes only modes that are symmetric under reflection in either the x-axis or the y-axis or both. In general, the representation might also include modes that are antisymmetric under reflection in the x-axis, the y-axis or both axes. However, antisymmetry with zero mean deflection implies zero deflection at the potential bond point. Consequently, the variance of the deflection of the membrane at the potential bond point due to fluctuations is independent of these antisymmetric modes and, therefore, these modes need not be taken into account.

The bonding potential

In order to describe adhesion, the interaction between the membrane and the substrate at prospective binding sites must be prescribed. Here, adhesion is represented by means of an interaction energy well, with the depth of the well corresponding to the energy reduction achieved when bonding occurs in the absence of other physical effects. In other words, it is the chemical potential of bonding. The width of the potential well represents the compliance of the bond. We believe this is a realistic description of bonding when the molecules involved in adhesion can be relatively large and compliant. As suggested by

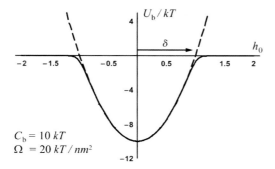

Fig. 2.12 The solid line illustrates the bonding potential defined in (2.27) for $C_b = 10\,kT$ and $\Omega = 20\,kT/nm^2$. The dashed line shows the parabolic approximation to the potential based on the depth at potential minimum and the curvature of the potential at that point.

Lin *et al.* [17], a convenient choice for the shape of such a well is

$$u_b(h_0) = -kT\ln\left[\left(e^{\frac{C_b}{kT}} - 1\right)e^{-\frac{\Omega}{2kT}h_0^2} + 1\right] \tag{2.27}$$

where u_b is the bonding interaction energy, h_0 is the deflection of the membrane in the z-direction at this binding site, and kT is the thermal energy unit; C_b is the depth of the energy well at deflection $h_0 = 0$ and Ω is the curvature of the well at the same point. Notice that C_b has the physical dimensions force $\times$ length. The physical units of Ω are force/length.

As has been pointed out in [17], the interaction potential shown in (2.27) can be approximated by a harmonic potential

$$u_b(h_0) \approx \frac{1}{2}\Omega h_0^2 - C_b \tag{2.28}$$

within the well itself. Figure 2.12 shows the comparison between the full potential and the harmonic approximation, where the solid line corresponds to (2.27) for $C_b = 10\,kT$ and $\Omega = 20\,kT/nm^2$; the dashed line represents the harmonic potential (2.28) for the same parameter values. As is evident from the figure, these two potentials are nearly identical when the deflection h_0 is deep in the well, and only when h_0 approaches the rim of the well does the quadratic potential begin to deviate significantly from the full potential. An estimate of the width of the well δ is given by (2.28) as

$$\delta = \sqrt{\frac{2C_b}{\Omega}}. \tag{2.29}$$

The well depth C_b and the well width δ are adopted as the system parameters which characterize the interaction energy between the centered adhesion receptor in the membrane and its compatible ligand attached to the substrate.

Stability of the molecular bond

The question of the stability of the bond between a centrally positioned molecule embedded within the membrane and its ligand attached to the substrate is now examined. Within the thermodynamic framework, we are able to form conclusions only on the probability that the bond has been completed at any instant or, equivalently, on the fraction of time that the bond is closed over relatively long periods of time. By stability of the bond in this instance, we mean only that the standard deviation σ of the bond point due to thermal fluctuations is no greater than a certain fraction of the width of the well, say $\sigma/\delta = 0.5$ or 0.2. If such a criterion is satisfied, the reaction state will be confined to the well strongly enough to consider the molecular pair to be bound.

According to classical statistical mechanics, the standard deviation $\sigma > 0$ of center point deflection due to fluctuations is given by the variance of the fluctuations at that point,

$$\sigma^2 = \frac{1}{Z} \int_{-\infty}^{\infty} \cdots \int_{-\infty}^{\infty} h(0,0)^2 e^{-(U_e+U_b)/kT} \, da_1 \cdots da_N \qquad (2.30)$$

where U_b is the interaction energy with profile $u_b(h_0)$ defined in (2.27) and a_k with $k = 1, \ldots, N$, $N = n_m^2$ is a one-dimensional array of random variables consisting of the elements of the matrix a_{ij} in any order. The denominator Z is the partition function (or configuration integral)

$$Z = \int_{-\infty}^{\infty} \cdots \int_{-\infty}^{\infty} e^{-(U_e+U_b)/kT} \, da_1 \cdots da_N. \qquad (2.31)$$

For values of C_b that are significantly greater than kT, the integrals defining the variance can be evaluated in closed form for quite large values of N. The feature that makes this possible is that the exponent of the Boltzmann factor in the integrand is a positive definite quadratic form in the random variables a_k, say $a_i Q_{ij} a_j$, where the Einstein summation convention is presumed. The $N \times N$ matrix Q_{ij} then has N real, positive eigenvalues, say $\mu^{(k)}$ for $k = 1, \ldots, N$, with due regard for multiplicity, and a corresponding set of N orthonormal eigenvectors $v_i^{(k)}$. The integration variables can then be replaced by a new set of variables, say b_k, according to $a_i = S_{ij} b_j$ where $S_{ij} = v_i^{(j)}$. As a result of this change in variables, the quadratic form reduces to $a_i Q_{ij} a_j = \mu^{(1)} b_1^2 + \cdots + \mu^{(N)} b_N^2$ and the variance then evaluates to

$$\frac{\sigma^2}{\delta^2} = \sum_{k=1}^{N} \frac{1}{2\mu^{(k)}} \left(\sum_{m=1}^{N} S_{km} \right)^2. \qquad (2.32)$$

Thus, the task of calculating the variance is reduced to that of finding the eigenvalues and eigenvectors of an $N \times N$ real, symmetric, positive definite matrix. In each case, the result can be cast in the nondimensional form

$$\frac{\sigma}{\delta} = F\left(\frac{\Delta}{\delta}, \frac{C_e}{kT}, \frac{C_b}{kT}\right) \tag{2.33}$$

which expresses the degree of confinement of the center point of the membrane within the interaction well, as represented by the nondimensional ratio σ/δ, in terms of nondimensional combinations of the remaining system parameters.

Among the system parameters appearing in (2.33), the one with values most thoroughly considered in the literature is the stiffness parameter C_e [16]. For lipid bilayer membranes, reported values fall in a range around the value $C_e = 30\,kT$, so this value is adopted for purposes of illustrating the implications of the result (2.33).

As a bonding criterion, we arbitrarily select $\sigma/\delta < 0.2$. The quality of the conclusions reached on this basis does not depend on the particular value chosen but, obviously, the range of states regarded as conducive to stable bond formation is narrowed (broadened) if the bonding criterion is made more (less) restrictive.

With the values of C_e/kT and σ/δ selected, the result (2.33) defines a curve in the plane of span length Δ/δ versus binding chemical potential C_b/kT. This curve is illustrated in Fig. 2.13 for $n_m = 11$ or, equivalently, 121 degrees of freedom. The interpretation of this graphical result is that, for systems with span lengths and interaction potential values falling below and to the right

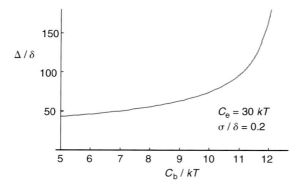

Fig. 2.13 For the parameter values shown and $n_m = 11$, the magnitude of the membrane fluctuation at the bond point is less than 0.2 for a system with parameter values that fall below and to the right of the curve, whereas the magnitude is greater than 0.2 for parameter values that lie above and to the left. The former is interpreted as a stable bond configuration, while the latter is unstable.

of the curve, connections are likely to remain bound once they are formed. Likewise, for systems with parameter values falling above and to the left of the curve, bonds are not likely to remain intact even though they may be formed with some frequency. The strength of the thermal fluctuations in such cases will invariably overcome any completed bond in the course of random fluctuations. This result establishes a clear connection between span length or binding site spacing and the tendency for molecular adhesions to form.

To extract a specific estimate of critical span length which is that value which distinguishes between stable and unstable bonds, adopt the value of $C_b/kT = 10$ which is based on experimental observations on an integrin-ligand bond reported in [18]. This implies a value of Δ/δ of about 70. For large adhesion proteins, we expect the bond to be fairly compliant. Adopting the value $\delta = 2\,\text{nm}$ to represent the width of a compliant bond well then implies a critical bond spacing of about 140 nm beyond which tight adhesion arrays can not form. This value is larger by a factor of about 2 than values observed in experiments reported by Arnold *et al.* [19] on the adhesion of several tissue secreting cell types to a hard substrate.

Finally, the dependence of the estimate of membrane fluctuation σ/δ at the bond point on the number of modes n_m used in the series approximation (2.25) is illustrated in Table 2.1 for a particular value of C_b. Values appear to be con-verging to a particular limit as n_w becomes relatively large, and the results for $n_m = 11$ already seem to provide a good approximation to the limiting value.

The influence of membrane tension

Up to this point in the discussion, the possibility that the membrane might be subjected to an in-plane tension has not been considered. However, it is possible that the patch of membrane being analyzed here is part of a cell or vesicle with an internal pressure, in which case the membrane would indeed be subjected to an in-plane tension. This is an *applied* force acting on the membrane that is capable of doing work as the membrane deflects from its ref-erence configuration, as opposed to an internal force arising as a consequence of deformation.

Denote this membrane tension by P, with units of force per unit length, and assume that it acts along the periphery of the membrane patch illustrated in Fig. 2.11 in the direction that is normal to the edge of the patch and in its reference plane.

The membrane is assumed to exhibit fluid-like in-plane behavior, so that the in-plane state of stress is equi-biaxial, and it is assumed to deform in such a way that the midplane area is conserved. As the membrane undergoes transverse deflection with local value $h(x, y)$, the actual area is unchanged but the projected

Table 2.1. *An illustration of the rate of*
convergence of the span length inferred from
(2.33) on the number of deflection modes n_m or
number of degrees of freedom $N = n_m^2$ in the
approximation (2.25). Use of a relatively small
number of modes leads to a good estimate of
the critical span length for a given value C_b of
bonding potential.

n_m	N	$\Delta/\delta\vert_{C_b=10\,kT}$
1	1	96.7
3	9	78.4
5	25	76.0
7	49	75.2
9	81	74.8
11	121	74.6
13	169	74.5
31	961	74.3

area in the x, y-plane must contract as a result of the deflection. This contraction works against the applied membrane tension P, thereby increasing the energy of the system. This increase in energy as a functional of the deflection is

$$U_p = P \int_{-\frac{\Delta}{2}}^{\frac{\Delta}{2}} \int_{-\frac{\Delta}{2}}^{\frac{\Delta}{2}} \left[(\partial_x h)^2 + (\partial_y h)^2\right] dx\,dy \qquad (2.34)$$

where the integral is the change in area of the patch, accurate to lowest order in the slope of the deformed membrane. The partition function is then defined as in (2.31) with the energy U_e replaced by $U_e + U_p$.

The question of stability of the molecular bond at the center of the membrane can be examined by following the steps outlined above. The calculations in the present instance are nearly identical to those for the case of bending resistance only, with the only difference being the inclusion of the quantity U_b/kT in the exponent of the Boltzmann factor.

The influence of the membrane tension is represented parametrically by means of the nondimensional quantity

$$\eta = \frac{P\delta^2}{kT}. \qquad (2.35)$$

Results of the calculation for $\eta = 0, 0.5, 1$, again with the elastic bending stiffness of the membrane chosen to be $C_e = 30\,kT$ and the criterion on the

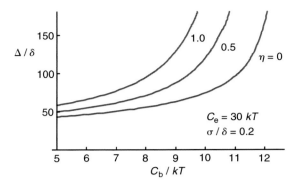

Fig. 2.14 The parameter η indicates the magnitude of the membrane tension in nondimensional form, and the curves define the boundary between regions of bond stability and bond instability (in the sense of Fig. 2.13) for three values of membrane tension and for $n_m = 11$. The curve labeled $\eta = 0$ is identical to the earlier result.

standard deviation of the fluctuations of the center of the membrane to be $\sigma/\delta \le 0.2$, are illustrated in Fig. 2.14. Once again, this is a graph of span length between bond points versus depth of the interaction energy well that separates the parameter range for which a completed bond is stable from the range for which it is not. The figure includes the obvious result that our earlier results are recovered for the case when $\eta = 0$. For the parameter values considered above, the dimensionless parameter $\eta = 1$ implies a membrane tension P of about 10^{-3} N/m. As expected, membrane tension has quite a strong stabilizing influence on bond stability.

Acknowledgments

This work is supported primarily by the MRSEC Program of the National Science Foundation at Brown University under award DMR-0520651.

References

[1]   Evans, E., Ritchie, K., and Merkel, R. Sensitive force technique to probe molecular adhesion and structural linkages at biological interfaces, *Biophysical Journal* **68** (1995) 2580–2587.

[2]   Evans, E. and Ritchie, K. Dynamic strength of molecular adhesion bonds, *Biophysical Journal* **72** (1997) 1541–1555.

[3]   Evans, E. A. and Calderwood, D. A. Forces and bond dynamics in cell adhesion, *Science* **316** (2007) 1148–1153.

[4]   Sotomayor, M. and Schulten, K. Single-molecule experiments in vitro and in silico, *Science* **316** (2007) 1144–1148.

[5]   Thomas, W. E. Understanding the counterintuitive phenomenon of catch bonds, *Current Nanoscience* **3** (2007) 63–83.

[6]   Berg, H. C. *Random Walks in Biology*, Princeton, NJ, Princeton University Press, 1993.
[7]   Risken, H. *The Fokker–Planck Equation*, Berlin, Springer-Verlag, 1989.
[8]   Hanngi, P., Talkner, P., and Morkovec, M. Reaction-rate theory: fifty years after Kramers, *Reviews of Modern Physics* **62** (1990) 251–342.
[9]   Heymann, B. and Grubmuller, H. Dynamic force spectroscopy of molecular adhesion bonds, *Physical Review Letters* **84** (2000) 6126–6129.
[10]   Thormann, E., Hansen, P. L., Simonsen, A. C., and Mouritsen, O. G. Dynamic force spectroscopy on soft molecular systems: improved analysis of unbinding spectra with varying linker compliance, *Colloids and Surfaces B – Biointerfaces* **53** (2006) 149–156.
[11]   Hanke, F. and Kreuzer, H. J. Breaking bonds in the atomic force microscope: theory and analysis, *Physical Review E* **74** (2006) 031 909.
[12]   Evans, E. Probing the relation between force lifetime and chemistry in singlemolecular bonds, *Annual Reviews of Biophysical and Biomolecular Structure* **30** (2001) 105–128.
[13]   Lin, Y. and Freund, L. B. A lower bound on receptor density for stable cell adhesion due to thermal undulations, *Journal of Materials Science* **42** (2007) 8904–8910.
[14]   Helfrich, W. Elastic properties of lipid bilayers: theory and possible experiments, *Zeitschrift für Naturforschung, Section C* **28** (1973) 693–703.
[15]   Seifert, U. Configurations of fluid membranes and vesicles, *Advances in Physics* **46** (1997) 13–137.
[16]   Boal, D. H. *Mechanics of the Cell*, Cambridge, Cambridge University Press, 2002.
[17]   Lin, Y., Inamdar, M., and Freund, L. B. The competition between Brownian motion and adhesion in soft materials, *Journal of the Mechanics and Physics of Solids* **55** (2007) 241–250.
[18]   Goennenwein, S., Tanaka, M., Hu, B., Moroder, L., and Sackmann, E. Functional incorporation of integrins into solid supported membranes on ultrathin films of cellulose: impact on adhesion, *Biophysical Journal* **85** (2003) 646–655.
[19]   Arnold, M., Cavalcanti-Adam, E. A., Glass, R., *et al.* Activation of integrin function by nanopatterned adhesive interfaces, *ChemPhysChem* **5** (2004) 383–388.

3

Statistical thermodynamics of cell–matrix interactions

TIANYI YANG AND MUHAMMAD H. ZAMAN

Introduction

Cells live, interact, and proliferate in a highly rich, diverse, and complex extracellular environment. Interactions with this dynamic extracellular environment occur through both signaling and adhesion receptors on the cell surface [1]. These receptors, which are typically transmembrane, are also responsible for regulating and modifying the cellular environment, just as the environment regulates the cell structure and function [2, 3]. In this chapter, we will focus primarily on the extracellular domain of the cell-adhesion receptors. This domain interacts with a variety of soluble and insoluble extracellular ligands. While these receptors are not limited to mechanical interaction and are essential for signaling, we will focus primarily on the physical and mechanical interactions between receptors and ligands with a discussion on implications for signaling presented in the final part of this chapter.

Mechano-chemical interactions of cell adhesion receptors play a phenomenal role in cellular function in both healthy and diseased cell state [4–6]. These receptors are responsible for adhesion, detachment and motility of a vast number of cells in vivo (Fig. 3.1). These processes are necessary for proper cell function, survival, and proliferation. At the same time, extracellular matrix not only affects this interaction, but also interacts directly with the cells through a number of adhesion receptors. Up- or downregulation of these receptors has been associated with increased invasion rates in a number of cancers with high incidence-to-death ratio [6]. Altered adhesion, due to either increase in

Statistical Mechanics of Cellular Systems and Processes, ed. Muhammad H. Zaman. Published by Cambridge University Press. © Cambridge University Press 2009.

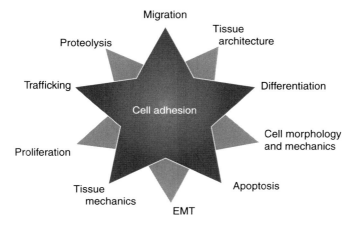

Fig. 3.1 Cell adhesion regulates numerous biological processes in vivo. A sample of some of these processes is depicted in the figure above. These processes range from maintenance of proper homeostasis (such as morphology, proliferation, trafficking) as well as diseases such as cancer, where events such as epithelial-to-mesenchymal transition (EMT) has been associated with cancer progression [7].

the adhesion-receptor concentration or loss of specific cell adhesion to other cells, has been associated with cancer progression and metastasis and is being studied in various cancer contexts. Cell–matrix adhesion is also the focus of a number of areas in biotechnology such as drug delivery, tissue engineering, and stem-cell therapy.

The interactions of cell surface receptors with motile and immobilized ligands are cooperative, coordinated, and complex. The cell surface receptors, that serve as primary anchors for cells to bind to the matrix also contribute in cell–cell interactions, though they are not the main contributor in those processes. Thus a fundamental level theoretical treatment of cell–matrix interactions is necessary for developing a basic understanding that has implications in cell biology, drug design, and healthcare.

Due to their enormous importance and fundamental nature, cell–matrix interactions have been studied in detail for decades [1, 2, 8–18]. While most of these efforts have been experimental in nature, theoretical endeavors have also studied these processes [19–35]. A majority of theoretical and computational tools have, however, utilized continuum-level mechanics, with little molecular-level detail. This is in part due to the nonequilibrium nature of many cell biological processes such as migration, adhesion under shear, etc. Additionally, the lack of fundamental-level theory on equilibrium processes in cell adhesion along with high computational costs of molecular simulations, have barred many researchers from attacking these problems. Nonetheless multiple

efforts over the last several decades have provided both qualitative and quantitative understanding of cell–matrix interactions, particularly for cell adhesion
in a variety of extracellular environments.

Among cell–matrix interactions, cell adhesion has been the topic of interest
in the majority of the computational studies aimed at elucidating biophysical and biomechanical mechanisms underlying adhesion. These models have
addressed a number of key problems in cell adhesion, ranging from leukocyte
rolling to cell spreading. The computational models can be broadly categorized
in the following main categories, based upon the length-scale of the particular
process:

(a)   Models at the molecular scale have used molecular dynamics or Monte
        Carlo type approaches, which are inherently stochastic and for most
        cases need atomistic-level resolution of receptors and ligands [35–38].
        While these models have been instrumental in elucidating how individ-
        ual receptors interact with ligands, their application has been limited
        for two main reasons. First of all, very few crystal structures at high
        resolution are available. As a result, these studies have been limited to
        a very small sample of receptors and ligands. Second, due to the enor-
        mous cost of running these simulations, a large number of receptors
        and interaction among them can not be studied.

(b)   Models aimed at cellular-level length-scale have primarily used contin-
        uum mechanics approaches and have addressed a number of nonequi-
        librium processes [19–33]. Like their single-molecule counterparts,
        these models have been successful in predicting effects at cellular
        length-scales. Issues such as leukocyte rolling, cell adhesion, and
        mechanics of cells have been analyzed in detail using these models;
        however, molecular-level events, entropic contributions from confor-
        mations, and effects of solvent are outside the realm of these models.

(c)   The final category of models have utilized statistical mechanics based
        strategies (equilibrium and nonequilibrium) to study bulk behavior.
        Since this is most closely related to the focus of this chapter, we will
        briefly review some of the efforts made in this category. Zhu has stud-
        ied the biophysics of cell adhesion interactions at the level of individual
        molecular pairs [39]. Application of a master-equation method for the
        kinetics of a receptor–ligand system in this model along with Monte
        Carlo simulations and constitutive equations coupling chemical reac-
        tions and mechanics describe several interesting biophysical aspects
        of cell–matrix interactions such as competitive binding between solu-
        ble and surface-bound ligands, lifetime adhesion under constant force,

detachment by ramp force, binding rate and rolling velocity, long-term evolution of adhesion, etc. Mogilner, Oster and co-workers [40–43] have developed nonequilibrium statistical mechanics models that capture the essential features of migration and force generation and include aspects of both classical and statistical mechanics. Shen and Wolynes [44] have developed nonequilibrium statistical mechanics models to study transegrity and cytoskeletal assembly. Finally, Bruinsma [45] has developed a model rooted in Langevin dynamics to study force generation by adhesion sites and has categorized three major stages in force generation: (1) initial adhesions (IA), (2) focal complexes (FC), and (3) focal adhesions (FA). FA are large complex structures, arising from FC and play an important role in cell signaling. The first two stages together are referred to as *nascent adhesion sites*. The question how dynamic adhesion sites can be activated by substrate rigidity is addressed.

Collectively taken, these efforts rooted in statistical mechanics and thermodynamics represent a very small fraction of the number of studies on cell adhesion. A significant number of studies have assumed adhesive substrates interacting with cells with little information on the molecular architecture of ligands. Effects of immobilization of ligands have also not gotten due attention. Thus a comprehensive picture of cell–matrix interactions, one that is able to account for molecular detail and predict bulk behavior is far from complete.

In the following sections, we describe our approach, rooted in previous studies of polymer adhesion and gel formation [46–48] to study cell–matrix interactions and dynamics of receptors on the cell surface during these processes.

Model system

We apply a statistical thermodynamic model to study cell-substrate adhesion through receptor–ligand interaction at molecular level, taking into consideration entropic, conformational, solvation, long-term, and short-term interactions between species. We describe free-energy landscape as a function of a number of experimentally measurable parameters, e.g., surface coverage of receptors or ligands, cell–substrate separation, conformational restriction on receptor, and interaction strength [34].

Our system of interest is a two-dimensional cell interacting with a rigid substrate decorated with ligands (Fig. 3.2). This system is based upon numerous adhesion, migration and cell–matrix interaction studies studied routinely in vitro. Our system consists of receptors and ligands with solvent effects modeled implicitly.

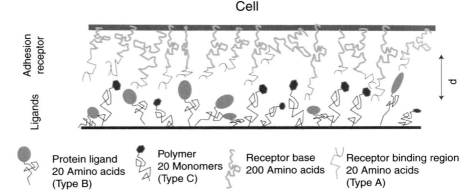

Fig. 3.2 A schematic of our cell–matrix adhesion model. The two-dimensional system includes cell membrane decorated with adhesion receptors and interactions between adhesion receptors and ligands of various types. The choice and the types of ligands can be extended to capture various scenarios observed in experiments. (Figure adapted from Yang *et al.*, *J. Chem. Phys.* **126**, 2007, with permission.)

Cell-matrix interaction with immobilized ligands

The first case of interest is a cell–matrix system with immobilized ligands. This system also mimics experiments carried out routinely where protein and polymer ligands are coated on the cell surface while controlling nontethered ligand concentration in the system.

The system of interest consists of three types of chain molecules with receptor (type A) tethered to the cell surface, matrix ligand protein (type B) and synthetic polymer (type C) attached to the substrate (Fig. 3.2). Besides chain molecules, the rest of the space between the two planes is filled with solvent. The area of either plane is A. The number of molecules A, or B, or C tethered, is N_A, N_B, or N_C respectively. Surface coverage for A, is $\sigma = N_A / A$, and the ratio of number of B or C to A is r_B and r_C, respectively. The separation between the two parallel planes is d. The system is homogeneous in $x - y$ planes and varies only with z. We use coarse-grain simulation to generate conformations of protein with dihedral angles chosen randomly from the sterically allowed regions (e.g., α, β basins, etc.) of the $\phi - \psi$ map. The side chains are simulated using noninteracting hard spheres. The polymer molecules are simulated with a similar strategy with three possible conformations, namely trans, gauche$^+$ and gauche$^-$.

For simplicity, we only consider interactions between different molecules, i.e. such as receptor–protein ligand (A–B), receptor–polymer (A–C), or protein ligand and polymer (B–C). Self-interaction terms are ignored. The interaction term contains an attractive and a repulsive part. We introduce the incompressibility condition, or volume-filling constraint to account for the short-term repulsive interaction. The constraint equation, in the layer $(z, z + dz)$ is:

$$\langle \varphi_A (z) \rangle + \langle \varphi_B (z) \rangle + \langle \varphi_C (z) \rangle + \varphi_S (z) = 1 \tag{3.1}$$

in which, $\langle \rangle$ denotes ensemble average, $\varphi(z)$ is volume fraction of according molecule in the layer $(z, z + dz)$. Subscripts A, B, C, and S denote molecules of receptors (A), protein on the substrate surface (B), polymer molecules on substrate (C), and solvent (S), respectively. The volume fraction can be expressed as:

$$\langle \varphi_A (z) \rangle = \frac{N_A \langle n_A (z) \rangle dz v}{A dz} = \sigma \sum_{\{\alpha\}} P_{A\alpha} \, n_{A\alpha} (z) \, v \tag{3.2}$$

where $\sigma = N_A / A$, is the surface coverage of receptor. $\langle n_A (z) \rangle dz$ is the number of segments of molecule A in the layer $(z, z + dz)$ averaged by the ensemble of conformations of molecule A. $n_{A\alpha} (z)$ is the density of segments of the α conformation of A in $(z, z + dz)$ layer. $P_{A\alpha}$ is the probability distribution function (pdf) of molecule A. Assuming the volume of molecule A, B, C, and solvent to be identical the constraint equation can be written as:

$$\sigma \sum_{\{\alpha\}} P_{A\alpha} \, n_{A\alpha} (z) \, v + r_B \, \sigma \sum_{\{\beta\}} P_{B\beta} \, n_{B\beta} (z) \, v + r_C \, \sigma \sum_{\{\gamma\}} P_{C\gamma} \, n_{C\gamma} (z) \, v + \varphi_S (z) = 1 \tag{3.3}$$

The free energy of the system has the following form:

$$f[\varphi_S (z), P_{A\alpha}, P_{B\beta}, P_{C\gamma}] = \frac{F}{kTA} = \sigma \sum_{\{\alpha\}} P_{A\alpha} \ln P_{A\alpha} + r_B \, \sigma \sum_{\{\beta\}} P_{B\beta} \ln P_{B\beta} + r_C \, \sigma \sum_{\{\gamma\}} P_{C\gamma} \ln P_{C\gamma}$$

$$+ r_B \, \sigma^2 \sum_{\{\alpha\}} \sum_{\{\beta\}} \iint dz dz' \, \chi_{AB} \, (|z - z'|) \, P_{A\alpha} \, P_{B\beta} \, n_{A\alpha} (z) \, n_{B\beta} (z')$$

$$+ r_B \, r_C \, \sigma^2 \sum_{\{\beta\}} \sum_{\{\gamma\}} \iint dz dz' \, \chi_{BC} \, (|z - z'|) \, P_{B\beta} \, P_{C\gamma} \, n_{B\beta} (z) \, n_{C\gamma} (z')$$

$$+ r_C \, \sigma^2 \sum_{\{\gamma\}} \sum_{\{\alpha\}} \iint dz dz' \, \chi_{CA} \, (|z - z'|) \, P_{C\gamma} \, P_{A\alpha} \, n_{C\gamma} (z) \, n_{A\alpha} (z')$$

$$+ v^{-1} \int dz \, \varphi_S (z) \ln \varphi_S (z) \tag{3.4}$$

The first three terms are due to the conformational entropy of chain molecules. The last term is caused by translational entropy of solvent. The three terms in the middle are contributions from intermolecular interactions.

Transform the constraint equation by the multiplier function $\pi(z)$ (which is osmotic pressure divided by $k_B T$):

$$g[\varphi_S(z), P_{A\alpha}, P_{B\beta}; P_{C\gamma}] = \int dz \pi(z) v\sigma \left(\sum_{\{\alpha\}} P_{A\alpha} n_{A\alpha}(z) + r_B \sum_{\{\beta\}} P_{B\beta} n_{B\beta}(z) \right.$$

$$\left. + r_C \sum_{\{\gamma\}} P_{C\gamma} n_{C\gamma}(z) \right) + \int dz \, \varphi_S(z) \pi(z) - \int dz \pi(z) = 0.$$

$$(3.5)$$

We apply the functional derivatives of f and g with respect to $\varphi_S(z)$, plug into the following relation:

$$\frac{\delta f}{\delta \varphi_S(z)} + \frac{\delta g}{\delta \varphi_S(z)} = 0. \tag{3.6}$$

Similar calculation of the other three equations for $P_{A\alpha}$, $P_{B\beta}$ and $P_{C\gamma}$, gives us a close set of coupled nonlinear equations for $\varphi_S(z)$, $P_{A\alpha}$, $P_{B\beta}$, $P_{C\gamma}$, and $\pi(z)$, together with the original constraint equation:

$$\varphi_S(z) = e^{-v\pi(z)} \tag{3.7}$$

$$P_{A\alpha} = \frac{1}{g_A \, E_{A\alpha}} e^{-r_B \sum_{\{\beta\}} M_{\alpha\beta}^{AB} P_{B\beta}} e^{-r_C \sum_{\{\gamma\}} M_{\alpha\gamma}^{AC} P_{C\gamma}} \tag{3.8}$$

$$P_{B\beta} = \frac{1}{g_B \, E_{B\beta}} e^{-\sum_{\{\alpha\}} M_{\beta\alpha}^{BA} P_{A\alpha}} e^{-r_C \sum_{\{\gamma\}} M_{\beta\gamma}^{BC} P_{C\gamma}} \tag{3.9}$$

$$P_{C\gamma} = \frac{1}{g_C \, E_{C\gamma}} e^{-\sum_{\{\alpha\}} M_{\gamma\alpha}^{CA} P_{A\alpha}} e^{-r_B \sum_{\{\beta\}} M_{\gamma\beta}^{CB} P_{B\beta}} \tag{3.10}$$

Where g_X ($X = A$, B, or C) is a normalization factor, for example, g_A satisfies $\sum_{\{\alpha\}} P_{A\alpha} = 1$.

The intermolecular interaction term is calculated using:

$$E_{XY} \equiv e^{v \int dz \pi(z) n_{XY}(z)} \qquad (X \in \{A, B, C\}, Y \in \{\alpha, \beta, \gamma\}). \tag{3.11}$$

We define the interaction matrix, M, containing the interaction terms between A and B (or B and C or A and C) as:

$$M_{UV}^{ST} \equiv \sigma \iint dz dz' \, \chi_{ST}(|z - z'|) n_{SU}(z) n_{TV}(z') \quad S, T \in \{A, B, C\}, U, V \in \{\alpha, \beta, \gamma\}). \tag{3.12}$$

Matrices Ms are calculated by mean-field method assuming Van der Waals type interactions. Equations (3.7–3.10) are solved by iteration after the discretization of space [46–48]. After that, substituting these calculated quantities back into (3.4), we obtain the free energy.

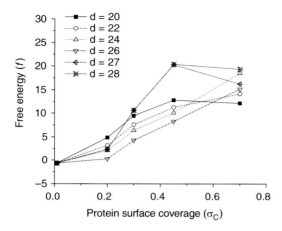

Fig. 3.3 Our model can quantify the effect of increase in surface coverage of the ligand (bottom plate in Fig. 3.2) on overall Helmholtz free energy of the system. The initial phase is dominated by attractive interactions, but as the distance changes and the concentration of the species increases the later phase is dominated by entropic repulsive interactions. (Figure reproduced from Yang *et al.*, *J. Chem. Phys.* **126**, 2007, with permission.)

Using a simple model our calculations provide a number of interesting features about cell–matrix interactions. First, we can study and characterize the free-energy landscape of adhesion in a noninteracting and an interacting system. Additionally, the effects of surface coverage can be studied (Fig. 3.3). These are particularly important as the surface coverage of receptors changes significantly during various stages of cancer progression.

Another interesting aspect of our calculation is our ability to study conformational confinement on cell–matrix adhesion. By restricting the conformations of the receptors to a single basin and comparing the results to the unrestricted conformations, we find that restriction on dihedral angles has a strong effect on free energy. Similarly sensitivity to other factors such as interaction potential or receptor and ligand separation as well as size can also be studied (Fig. 3.4).

The model discussed above uses coarse-grained, hard-sphere type simulations to generate conformations. However, more realistic potentials as well as structures to mimic amino acids making up the receptors and ligands can also be utilized. Since the crystal structure of many of these integrins is not yet available, trying to fold the entire sequence may not yield any productive results. On the other hand, simulating conformations of small fragments of the protein at a time, using a sliding window approach may result in more quantitative comparisons with bulk experiments. Since generation of conformations can be

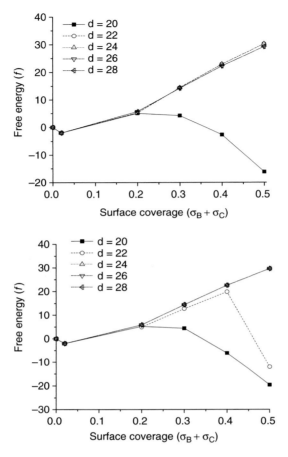

Fig. 3.4 In a mixed-species system (that includes both polymer and protein ligands tethered to the surface) shorter distances lead to decrease in free energy at high surface coverage due to attractive interactions between ligands and the receptors (top). This decrease in free energy is further amplified by conformational restrictions. Upon forcing the receptors to occupy only the β conformation, the overall free energy decreases, suggesting a strong dependence on conformational restrictions. (Figure reproduced from Yang *et al.*, *J. Chem. Phys.* **126**, 2007, with permission.)

achieved using parallel algorithms including GRID computing, such initiatives may result in studying the processes in greater detail.

Cell–matrix interaction with mobile ligands

Interactions of adhesion receptors in vivo and in vitro is not limited to immobilized ligands, but includes both motile and immobile ligands.

An extension of the statistical thermodynamic model to study interaction of receptors with motile ligands allows for quantitative study of cell–matrix interactions, including studying the effect of ligand density, receptor concentration, ligand size, and receptor–ligand interaction energy.

The model system for our motile ligand system is similar to Fig. 3.1 except for the presence of mobile ligands instead of immobilized tethered species. Ligands are modeled as spherical nano particles. Solvent molecules are imbued in the space between the membrane and the substrate. The system represents a canonical ensemble, with constant temperature, volume, and particle number where at equilibrium stage the Helmholtz free energy is minimized. The free-energy calculation of our system is somewhat similar to the tethered ligand case. The free energy can be separated into two categories, the entropy-related terms and interaction terms. The interaction terms are similar to the tethered system while there are significant differences in the entropy-related component. The entropy of our system can be separated into the following terms: (1) the translational or mixing entropy of solvent S_S, (2) the mixing entropy of spherical nano ligands S_E, (3) the translational entropy of receptor (including both bound and unbound) on the cell's membrane S_R, (4) the mixing entropy of free and bound receptors on the cell's membrane S_{F-B}, and (5) the conformational entropy of both free and bound receptors S_C.

$$S_S = -\frac{S}{V_S} \int dz\, \varphi_S(z) \log \varphi_S(z) \tag{3.13}$$

where V_S is the Van der Waals volume of solvent, which is assumed to be identical to the volume of one segment of receptor, S is the cross-section area of either plate, and $\phi_S(z)$ is the volume fraction of solvent at z.

$$S_E = -S \int dz\, \rho_E(z) \log \rho_E(z)\, V_E \tag{3.14}$$

where $\varphi_S(z)$ is the volume fraction of extracellular matrix (ECM) particle at position z, V_E is the volume of an ECM particle, $\rho_E(z)$ is the volume density for ECM particles at position z.

$$S_R = -N_R \log \varphi_R \tag{3.15}$$

where N_R is the total number of receptors, and φ_R is the area fraction of receptor molecules that cover the cell's membrane.

$$S_{F-B} = -N_R f_F \log f_F - N_R f_B \log f_B \tag{3.16}$$

where f_F is the fraction of free receptors, so $N_R f_F$ is the total number of free receptors. And f_B is the fraction of bound receptors.

$$S_C = -N_R f_F \sum_{\{\alpha\}} P_\alpha \log P_\alpha - N_R f_B \sum_{\{\beta\}} P_\beta \log P_\beta \tag{3.17}$$

where P_α is the probability of finding the α conformation of free receptor. P_β is probability of finding the β conformation of bound receptor. The Boltzmann constant k_B is taken to be 1 in the above equations. As shown by recent studies by Szleifer and co-workers [49] as well as Wang and co-workers [50] an explicit treatment of receptor–ligand interaction binding is critical for quantitative description of receptor–ligand interactions. The explicit interaction I_E refers to bound energy with respect to receptor–ligand. For the implicit interaction I_I we'll use Lennard–Jones type potential. Here R denotes receptor and E denotes nano-sized ligand. For simplicity, we assume their interaction strength to be all the same. The explicit interaction can be described as:

$$I_E = -N_R f_B \, \varepsilon_0 \tag{3.18}$$

where ε_0 is the magnitude of binding energy of one receptor–ligand bond (in the range of 10–$30 \, kT$). Similarly, the implicit interaction can be described by

$$
\begin{aligned}
I_I = &-\frac{s^2}{2 v_S^2} \iint d z_1 \, d z_2 \, \chi'_{11} \, (|z_1 - z_2|) \, [1 - \varphi_S (z_1)] \, [1 - \varphi_S (z_2)] \\
&-\frac{s^2}{v_S^2} \iint d z_1 \, d z_2 \, \chi'_{12} \, (|z_1 - z_2|) \, [1 - \varphi_S (z_1)] \, \varphi_S (z_2) \\
&-\frac{s^2}{2 v_S^2} \iint d z_1 \, d z_2 \, \chi'_{22} \, (|z_1 - z_2|) \, \varphi_S (z_1) \, \varphi_S (z_2)
\end{aligned}
\tag{3.19}
$$

where $\chi'_{ij} (|z_1 - z_2|)$ is the implicit interaction strength between polymers, polymer and solvent, solvent and solvent, respectively. Here polymer includes both receptor and the ECM particle. We define a dimensionless $\chi_{ij} = \beta \chi'_{ij}$. The free energy term, f, is similar to the term described in the case of immobilized ligands, but includes the entropic contributions unique to motile ligands.

$$
\begin{aligned}
f = &\, \beta A / S \\
= &\, \frac{1}{V_S} \int d z \, \varphi_S (z) \log \varphi_S (z) + \int d z \, \rho_E (z) \log \rho_E (z) \, V_E \\
&+ \sigma_R \log (s \, \sigma_R) + \sigma_R f_F \log f_F + \sigma_R f_B \log f_B \\
&+ \sigma_R f_F \sum_{\{\alpha\}} P_\alpha \log P_\alpha + \sigma_R f_B \sum_{\{\beta\}} P_\beta \log P_\beta \\
&- \sigma_R f_B \, \varepsilon_0 \\
&- \frac{S}{2 v_S^2} \iint d z_1 \, d z_2 \, \chi_{11} \, (|z_1 - z_2|) \, [1 - \varphi_S (z_1)] \, [1 - \varphi_S (z_2)]
\end{aligned}
$$

$$-\frac{S}{V_S^2} \iint d z_1 \, d z_2 \, \chi_{12} \left(|z_1 - z_2|\right) \left[1 - \varphi_S \left(z_1\right)\right] \varphi_S \left(z_2\right)$$

$$-\frac{S}{2 V_S^2} \iint d z_1 \, d z_2 \, \chi_{22} \left(|z_1 - z_2|\right) \varphi_S \left(z_1\right) \varphi_S \left(z_2\right) \tag{3.20}$$

where A is the free energy of the system, $\beta = 1/k_B T$, $\sigma_R = N_R/S$ is the surface coverage of receptors, and s is the area covered by one segment of receptor on the membrane.

The calculation of key thermodynamic parameters is similar to the discretiza-tion of space strategy used in the previous section. A natural consequence of our method is the chemical equilibrium constant K_{eq}:

$$K_{eq} = \frac{[R_B]}{[E_F][R_F]} = \frac{f_B}{f_F \left(r_E - f_B\right) [R_{tot}]} = \frac{f_B \, D}{f_F \left(r_E - f_B\right) \sigma_R} \tag{3.21}$$

in which D is the separation between the membrane and the substrate. In the bracket, R and E abbreviate for receptor and ECM, respectively. We focus on two key thermodynamic parameters to quantify the role of nano-ligands' size, concentration and interaction energy, namely the equilibrium constant K_{eq} and Helmholtz free energy f. The chemical equilibrium constant K is also associated with the equilibrium Gibbs free energy of interaction between the receptors and the ligands through $dG = -RT \ln K_{eq}$.

A number of interesting features emerge through our calculation. The most interesting and prominent is the behavior of the system at low and high con-centrations of ligands. For clarity, we keep the concentration ratio of receptors to ligands fixed at 5:1 (i.e., 5 ligands for each receptor). Figure 3.5 shows the behavior of a representative system (1.0 nm) as a function of concentra-tion (x-axis) and interaction energy. The interaction energy effect only scales the overall behavior of $-\ln K$ with the overall features staying intact. Simi-larly, the Helmholtz free energy of the system is unaffected by interaction energy. These results seem reasonable since variations in interaction energy typ-ically only become dominant if they induce conformational changes; however, since in this simple model there is no real conformational change associated with variations in interaction energy the primary effect observed is that of scaling.

The effects of size and concentration are shown in Fig. 3.6. At low concen-tration, we observe two fundamental trends. First of all, the $\ln K$ term is lower for bigger particles and decreases as the particle size goes down from 2.0 nm to 1.0 nm. The second key feature of the results is that concentration seems to have a stronger influence on bigger size nanoparticles as the slope of 2.0 nm is greater than that of 1.0 nm. These features are no longer present at higher

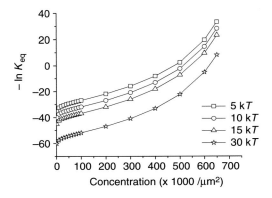

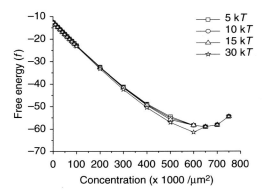

Fig. 3.5 The log equilibrium constant (ln K) and Helmholtz free energy show only a minor change with changes in overall interaction energy between receptors and ligands for a 1.0-nm particle. The effects are much more pronounced for effects due to ligand size (Fig. 3.6). (Figure reproduced from Yang and Zaman, *Langmuir* **24**, 2008, with permission.)

concentration. As a matter of fact, opposite trends are observed, as 1.5-nm and 2.0-nm systems show saturation while 1.0-nm system does not.

These effects, which can not be captured by simple scaling arguments from low concentration to high, suggest that nanoparticle size is of much greater consequence than interaction energy. They also provide insights into concentration effects that are important both in vivo and in the design of artificial systems. Above all, these methods provide a platform to develop a more detailed approach incorporating ligands of bigger radii, receptors with tunable flexibility and amino acid composition and systems that include both motile and tethered ligands. Such systems can be constructed by rather straightforward extrapolation of our models discussed in the previous sections and will provide interesting insights to both theoreticians and experimentalists.

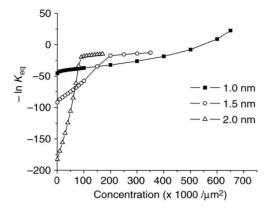

Fig. 3.6 The mobile ligand–receptor system is highly sensitive to ligand size, and leads to completely opposing trends at low and high concentrations. Increasing concentration leads to a phase transition from 2.0 to 1.0 being the preferred ligand size for lower Gibbs free energy. (Figure reproduced from Yang and Zaman, *Langmuir* **24**, 2008, with permission.)

Receptor clustering and implications for signaling

The previous two sections discuss the construction and implementation of a model system to describe cell–matrix adhesions in tethered and mobile ligand systems. While the models may lack some essential features, they provide a fundamental thermodynamic foundation that can be used to construct models that are more detailed and quantitative. These models also present a departure from numerous existing computational strategies that are unable to provide bulk level information or are limited to a single length-scale.

In this section, we turn our attention to a related issue that not only influences interaction of matrices with cellular receptors but also regulates the signaling processes that follow adhesion and interactions between receptors and ligands. This process is clustering of adhesion receptors on the cell surface as they come in contact with outside ligands. Clustering can also occur in the absence or lower concentration of ligands, but is stabilized by external ligands, which may be immobilized or motile. Clustering of integrins on the cell surface leads to focal complex and focal adhesion formation and hence provides a good metric to compare with experiments and large-scale simulations.

While clustering has gained significant attention from the continuum mechanics community as of late [51], including models that are able to capture numerous experimental scenarios, the interface between clustering and adhesion thermodynamics has remained elusive. As the focus of this chapter is statistical thermodynamic description of cell–matrix interactions, we will address clustering in this context, and not in the context of force generation or response to external mechanical stimuli.

To interface between cell adhesion statistical thermodynamics and clustering of receptors, we use a two-step approach, with the first step being the calculation of equilibrium cluster populations and the second step being the calculation of free energy of adhesion. Our system of interest is essentially identical to Fig. 3.1, with the exception being that ligands are now able to engage multiple receptors rather than a single receptor. While cluster formation can be directed by ligands or can happen through diffusion, we discuss a simple case here where first an equilibrated cluster population is achieved and then it interacts with the ligands. The other, more complicated case can be extended from the current description.

The equilibrium population is calculated using a Monte Carlo approach, which is a modification of the model used before by Linderman and co-workers [52]. Receptor conformations are generated in a similar manner to other models described in this chapter with explicit treatment of the backbone and hard-sphere models for the side chains. Clustering or dimerization of receptors is allowed only when the receptors are in the correct conformation and within a given distance of their partner. Receptors can diffuse on the cell surface whether the cell membrane is flat or curved. The diffusivities of the receptors are a function of receptor size, diameter and the properties of the membrane tethering them.

As the receptors diffuse on the cell surface, they come in contact with other receptors. The interaction of receptors with other receptors leads to dimerization, and eventually to cluster formation. Classical interaction potentials describe interaction between receptors, with the energetic advantage of dimerization greater than dissociation or monomerization. The excluded volume constraint is executed by the calculation of center of masses of the receptors and the interacting regions to ensure that steric constraints of binding are satisfied. Since there is a finite probability of dissociation and monomerization, over long times, equilibrium is established where the cluster population remains the same within tolerance. At this point, we freeze the conformations and distributions of receptors in a cluster and allow it to interact with ligands in the opposing plane using methods described previously. By varying the initial population of the receptor, interaction energy, and receptor and ligand composition and size, this approach allows to study the energetic advantage of clustering.

Our results indicate that clustering is only energetically beneficial if the initial concentration of receptors is high and the interaction energy between receptors and ligands is also significant (Fig. 3.7). In other words, clustering provides no benefit in terms of the free energy of the system if the ligand–receptor

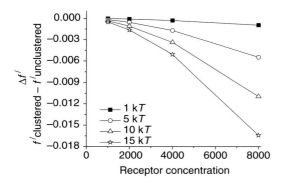

Fig. 3.7 The difference between the free energy of clustered and unclustered systems suggests that clustering is only beneficial at high initial concentration of receptors coupled with higher interaction energies between receptors and ligands. (Figure reproduced from Yang and Zaman, *Chem. Phys. Lett.* **454**, 2008, with permission.)

interactions are weak. Thus, in experimental systems that utilize polymers as substrates, with poor or weak binding to integrins and focal contacts, we predict that the cluster size will be small and the energetic advantage of clustering to the cell will be minimal.

While this is a simple model, it provides a potentially useful bridge between cell-surface dynamics and interactions of receptors with ligands. Incorporation of motile ligands is an obvious and a straightforward extension. Similarly, membrane curvature and ligand composition can also provide useful results. But perhaps the most interesting results will be the construction of similar interfaces between clusters and cell signaling. It is well known that cell signaling through integrins depends strongly on the ability of integrins to form clusters and the link between cluster size and signaling ability of cells remains unclear. Experiments and simulations to probe any possible link between signaling and cluster size, cluster stability, and strength of interactions between clusters and their ligands will undoubtedly be a major step in bridging the gap between cell biology and statistical mechanical approaches to study cell–matrix interactions. It will also be a critical piece in developing a single multiscale framework to model cellular events from extracellular mechanics, composition, and properties to cell signaling.

Concluding remarks and future directions

Cell–matrix interactions represent a large family of events and processes that regulate cell fate and function. With implications from immunology to biotechnology, tissue engineering to cancer, understanding cell–matrix interactions at the fundamental thermodynamic level is critical in both basic and applied sciences. The models and approaches presented in this chapter are just some of the approaches being used currently to understand both equilibrium and nonequilibrium processes involved in cell–matrix interactions. Significant challenges still remain to be overcome before these models can be used for quantitative predictions and design of novel experiments. Among these challenges, multiscaling of the processes from single molecular events to cellular-level processes is a key road block. The models presented here have the potential to overcome this challenge and efforts are under way to make these approaches fully multiscale. We hope that teams of researchers, bringing together experimentalists and theoreticians, will develop other models, rooted in fundamentals of statistical mechanics to address these challenges in multiscale modeling, both in the spatial and in the temporal regimes.

Other issues that need our immediate attention are incorporation of nonequilibrium events such as shear forces, dynamic chemical gradients, and motility of cells under the influence or absence of external forces. The models presented here deal exclusively with equilibrium and are not able to capture the nonequilibrium statistical mechanics of these processes. Integration of these models with continuum approaches along with more biological input would also be critical for their widespread applicability to a large number of biological problems.

In spite of these limitations, theoretical models presented in this chapter represent a new stream of statistical thermodynamic approaches to study fundamental processes in cell biology. Continuation of these efforts, together with experiments, will not only enhance our ability to create more accurate and comprehensive models but will also provide useful information to industry and the clinic in designing substrates, interfaces, and scaffolds to control and regulate cell–matrix interactions.

Acknowledgments

MHZ would like to acknowledge the gracious support of the Robert A. Welch Foundation (Grant F-1677).

References

1. Hynes, R. O., Integrins: bidirectional, allosteric signaling machines. *Cell*, 2002. **110**(6): 673–87.
2. DeSimone, D. W., M. A. Stepp, R. S. Patel, and R. O. Hynes, The integrin family of cell surface receptors. *Biochem Soc Trans*, 1987. **15**(5): 789–91.
3. Hynes, R. O., Integrins: a family of cell surface receptors. *Cell*, 1987. **48**(4): 549–54.
4. Critchley, D. R., Focal adhesions: the cytoskeletal connection. *Curr Opin Cell Biol*, 2000. **12**(1): 133–9.

5.   Li, S., J. L. Guan, and S. Chien, Biochemistry and biomechanics of cell motility. *Annu Rev Biomed Eng*, 2005. **7**: 105–50.

6.   Hynes, R. O., J. C. Lively, J. H. McCarty, *et al.*, The diverse roles of integrins and their ligands in angiogenesis. *Cold Spring Harb Symp Quant Biol*, 2002. **67**: 143–53.

7.   Hanahan, D. and R. A. Weinberg, The hallmarks of cancer. *Cell*, 2000. **100**(1): 57–70.

8.   Friedl, P., E. B. Brocker, and K. S. Zanker, Integrins, cell matrix interactions and cell migration strategies: fundamental differences in leukocytes and tumor cells. *Cell Adhes Commun*, 1998. **6**(2–3): 225–36.

9.   Hynes, R. O. and Q. Zhao, The evolution of cell adhesion. *J Cell Biol*, 2000. **150**(2): F89–96.

10. Francis, S. E., K. L. Goh, K. Hodivala-Dilke, *et al.*, Central roles of alpha5beta1 integrin and fibronectin in vascular development in mouse embryos and embryoid bodies. *Arterioscler Thromb Vasc Biol*, 2002. **22**(6): 927–33.

11. Hynes, R. O., The emergence of integrins: a personal and historical perspective. *Matrix Biol*, 2004. **23**(6): 333–40.

12. Cukierman, E., R. Pankov, D. R. Stevens, and K. M. Yamada, Taking cell-matrix adhesions to the third dimension. *Science*, 2001. **294**(5547): 1708–12.

13. Akiyama, S. K. and K. M. Yamada, The interaction of plasma fibronectin with fibroblastic cells in suspension. *J Biol Chem*, 1985. **260**(7): 4492–500.

14. Akiyama, S. K., E. Hasegawa, T. Hasegawa, and K. M. Yamada, The interaction of fibronectin fragments with fibroblastic cells. *J Biol Chem*, 1985. **260**(24): 13 256–60.

15. Cukierman, E., R. Pankov, and K. M. Yamada, Cell interactions with three-dimensional matrices. *Curr Opin Cell Biol*, 2002. **14**(5): 633–9.

16. Humphries, M. J., K. Olden, and K. M. Yamada, A synthetic peptide from fibronectin inhibits experimental metastasis of murine melanoma cells. *Science*, 1986. **233**(4762): 467–70.

17. Even-Ram, S. and K. M. Yamada, Cell migration in 3D matrix. *Curr Opin Cell Biol*, 2005. **17**(5): 524–32.

18. Zamir, E., M. Katz, Y. Posen, *et al.*, Dynamics and segregation of cell-matrix adhesions in cultured fibroblasts. *Nat Cell Biol*, 2000. **2**(4): 191–6.

19. Hammer, D. A. and D. A. Lauffenburger, A dynamical model for receptor-mediated cell adhesion to surfaces. *Biophys J*, 1987. **52**(3): 475–87.

20. Hammer, D. A. and D. A. Lauffenburger, A dynamical model for receptor-mediated cell adhesion to surfaces in viscous shear flow. *Cell Biophys*, 1989. **14**(2): 139–73.

21. Hammer, D. A. and S. M. Apte, Simulation of cell rolling and adhesion on surfaces in shear flow: general results and analysis of selectin-mediated neutrophil adhesion. *Biophys J*, 1992. **63**(1): 35–57.

22. Hammer, D. A., L. A. Tempelman, and S. M. Apte, Statistics of cell adhesion under hydrodynamic flow: simulation and experiment. *Blood Cells*, 1993. **19**(2): 261–75; discussion 275–7.

23. Ward, M. D., M. Dembo, and D. A. Hammer, Kinetics of cell detachment: peeling of discrete receptor clusters. *Biophys J*, 1994. **67**(6): 2522–34.

24. Ward, M. D., M. Dembo, and D. A. Hammer, Kinetics of cell detachment: effect of ligand density. *Ann Biomed Eng*, 1995. **23**(3): 322–31.

25. Kuo, S.C. and D.A. Lauffenburger, Relationship between receptor/ligand binding affinity and adhesion strength. *Biophys J*, 1993. **65**(5): 2191–200.

26. Alt, W. and M. Dembo, Cytoplasm dynamics and cell motion: two-phase flow models. *Math Biosci*, 1999. **156**(1–2): 207–28.

27. Dembo, M., D.C. Torney, K. Saxman, and D. Hammer, The reaction-limited kinetics of membrane-to-surface adhesion and detachment. *Proc R Soc Lond B Biol Sci*, 1988. **234**(1274): 55–83.

28. Tranquillo, R.T. and D.A. Lauffenburger, Stochastic model of leukocyte chemosensory movement. *J Math Biol*, 1987. **25**(3): 229–62.

29. Lauffenburger, D., B. Farrell, R. Tranquillo, A. Kistler, and S. Zigmond, Gradient perception by neutrophil leucocytes, continued. *J Cell Sci*, 1987. **88** (4): 415–16.

30. Tranquillo, R.T., D.A. Lauffenburger, and S.H. Zigmond, A stochastic model for leukocyte random motility and chemotaxis based on receptor binding fluctuations. *J Cell Biol*, 1988. **106**(2): 303–9.

31. Moghe, P.V. and R.T. Tranquillo, Stochastic model of chemoattractant receptor dynamics in leukocyte chemosensory movement. *Bull Math Biol*, 1994. **56**(6): 1041–93.

32. Tranquillo, R.T. and W. Alt, Stochastic model of receptor-mediated cytomechanics and dynamic morphology of leukocytes. *J Math Biol*, 1996. **34**(4): 361–412.

33. DiMilla, P.A., K. Barbee, and D.A. Lauffenburger, Mathematical model for the effects of adhesion and mechanics on cell migration speed. *Biophys J*, 1991. **60**(1): 15–37.

34. Yang, T. and M.H. Zaman, Free energy landscape of receptor mediated cell adhesion. *J Chem Phys*, 2007, **126**(4): 045103.

35. Zaman, M.H., Understanding the molecular basis for differential binding of integrins to collagen and gelatin. *Biophys J*, 2007. **92**(2): L17–19.

36. Krammer, A., H. Lu, B. Isralewitz, K. Schulten, and V. Vogel, Forced unfolding of the fibronectin type III module reveals a tensile molecular recognition switch. *Proc Natl Acad Sci USA*, 1999. **96**(4): 1351–6.

37. Lu, H., A. Krammer, B. Isralewitz, V. Vogel, and K. Schulten, Computer modeling of force-induced titin domain unfolding. *Adv Exp Med Biol*, 2000. **481**: 143–60; discussion 161–2.

38. Lu, H. and K. Schulten, Steered molecular dynamics simulations of force-induced protein domain unfolding. *Proteins*, 1999. **35**(4): 453–63.

39. Zhu, C., Kinetics and mechanics of cell adhesion. *J Biomech*, 2000. **33**(1): 23–33.

40. Mogilner, A. and G. Oster, Polymer motors: pushing out the front and pulling up the back. *Curr Biol*, 2003. **13**(18): R721–33.

41. Mogilner, A. and G. Oster, Force generation by actin polymerization II: the elastic ratchet and tethered filaments. *Biophys J*, 2003. **84**(3): 1591–605.

42. Bottino, D., A. Mogilner, T. Roberts, M. Stewart, and G. Oster, How nematode sperm crawl. *J Cell Sci*, 2002. **115**(2): 367–84.

43. Mogilner, A. and G. Oster, Cell motility driven by actin polymerization. *Biophys J*, 1996. **71**(6): 3030–45.

44. Shen, T. and P.G. Wolynes, Nonequilibrium statistical mechanical models for cytoskeletal assembly: towards understanding tensegrity in cells. *Phys Rev E Stat Nonlin Soft Matter Phys*, 2005. **72**(4 Pt 1): 041927.

45.   Bruinsma, R., Theory of force regulation by nascent adhesion sites. *Biophys J*, 2005. **89**(1): 87–94.

46.   Huang, Y. B., I. Szleifer, and N. A. Peppas, A molecular theory of polymer gels. *Macromolecules*, 2002. **35**(4): 1373–80.

47.   Huang, Y. B., I. Szleifer, and N. A. Peppas, Gel–gel adhesion by tethered polymers. *J Chem Phys*, 2001. **114**(7): 3809–16.

48.   Szleifer, I. and M. A. Carignano, Tethered polymer layers. *Adv Chem Phys*, 1996. **94**: 165–260.

49.   Longo, G. and I. Szleifer, Ligand–receptor interactions in tethered polymer layers. *Langmuir*, 2005. **21**(24): 11 342–51.

50.   Martin, J. I., C. Z. Zhang, and Z. G. Wang, Polymer-tethered ligand-receptor interactions between surfaces. *J Polymer Sci B Polymer Phys*, 2006. **44**(18): 2621–37.

51.   Qian, J., J. Wang, and H. Gao, Lifetime and strength of adhesive molecular bond clusters between elastic media. *Langmuir*, 2008. **24**(4): 1262–70.

52.   Brinkerhoff, C. J. and J. J. Linderman, Integrin dimerization and ligand organization: key components in integrin clustering for cell adhesion. *Tissue Eng*, 2005. **11**(5–6): 865–76.

4

Potential landscape theory of cellular networks

JIN WANG

Potential landscape and cellular network

To understand the biological function and robustness of the biological networks, it is crucial to uncover the underlying global principle [1-3]. The natures of the biological network have been explored by many experimental techniques [4], due to recent advances in genomics and proteomics. It has been found that biological networks are in general quite robust against genetic and environmental perturbations. Our goal is to understand this.

Bioinformatics studies

There are increasing numbers of studies on the global topological structures of the networks recently from bioinformatics [5-8] as well as from engineering design perspectives [9]. The scale-free properties and architectures for the networks have been elucidated [6-8]. The highly connected nodes in the network essential in keeping the network together might be important for the network's function. However, there have been so far very few studies of why the network should be robust and perform the biological function from the physical point of view [10-21].

Deterministic physical approaches

Theoretical models of the biological networks have often been formulated with a set of deterministic evolution equations either in the Boolean or continuous form [22-25, 34-37]. For cellular networks, they are chemical rate equations. These models are fruitful in exploring the local properties such as

Statistical Mechanics of Cellular Systems and Processes, ed. Muhammad H. Zaman. Published by Cambridge University Press. © Cambridge University Press 2009.

local stability of the network, but hard for the global ones. The dynamics of many networks at different levels can often be described with the general form of ordinary differential equations: $\dot{\mathbf{C}} = \mathbf{F}(\mathbf{C})$ with $\mathbf{C}$ as a concentration vector variable and each component of which representing the concentration of each species involved, and $\mathbf{F}$ as being the rate flux vector or driving (drifting) force. In general, one can not write the right-hand side of these equations as the gradient of a potential energy function. Since the potential function often can not be found, global properties such as stability and function of the whole system are therefore not easy to address in this approach.

Stochastic approaches

In reality, fluctuations are unavoidable. Statistical fluctuations coming from the finite number of basic building blocks (molecules for cellular networks, neurons for neuron networks and species in ecology networks) provide the source of intrinsic internal noise and the fluctuations from highly dynamical and inhomogeneous environments around provide the source of the external noise for the networks [25, 26, 28–33]. Both internal and external fluctuations play important roles in determining the nature of the network.

To take into account the underlying statistical fluctuations, efforts have been made to study the biological networks through a set of stochastic ordinary differential equations [25, 26] or kinetic Monte Carlo method either in Boolean or continuous form [27]. These stochastic descriptions are important and useful but inherently local given the limited amount of search time available. To probe the global properties, one has to explore the whole state space. The state space is typically cosmologically big (the number of the states). The global properties therefore are hard to see from this approach.

Landscape framework of biological networks

Here we will explore the global principles of the nonequilibrium biological networks from a different angle than the previous studies. We will formulate the network problem in terms of the potential function or potential landscape. If the potential landscape of the cellular network is known, the global properties can be explored [10, 12–21, 38, 39]. This is in analogy with the fact that in equilibrium statistical mechanics, the global thermodynamic properties can be explored when knowing the inherent interaction potentials in the system.

It is the purpose of this review to summarize some of the recent studies on the global robustness problem directly from the properties of the potential landscape. Furthermore, a cellular network is an open nonequilibrium system due to the interactions with the environment. There is often a dissipation cost associated with the network. It will also beinteresting to see how the dissipation cost is related to the features of the landscape reflecting the stability and robustness of the network.

We are going to uncover the underlying potential landscapes of a few cellular networks: the MAPK signal transduction network, the budding yeast cell cycle network, and the toggle switch of the gene regulatory network.

MAPK signal transduction

To explore the nature of the underlying potential landscape of the cellular networks, we will study a relatively simple yet important example of MAP-kinase signal transduction network [17] (Fig. 4.1). Mitogen-activated protein kinases (MAPK) belong to a family of serine/threonine protein kinases widely conserved among eukaryotes and are involved in many cellular programs such as proliferation, differentiation, movement and cell death. MAPK signaling cascades are organized hierarchically into three-tiered modules. MAPKs are phosphorylated and activated by MAPK-kinases (MAPKKs), which in turn are phosphorylated and activated by MAPKK-kinases (MAPKKKs). The MAPKKK is

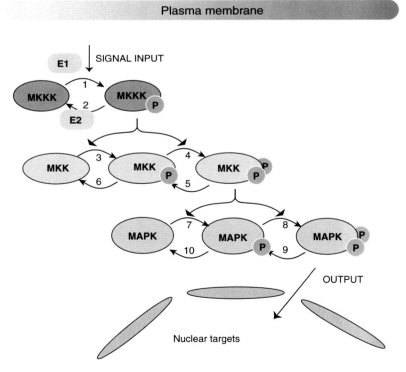

Fig. 4.1 The MAP-kinase network reaction scheme.

in turn activated by interaction with a family of small GTPases and/or other protein kinases connecting the MAPK module to the cell surface receptor or external stimuli [2, 3] (Fig. 4.1). We will study the global stability by exploring the underlying potential energy landscape for the MAP-kinase network.

Method

Let us study the global robustness of the network by starting from the network of chemical reactions in noisy fluctuating environments [17]:

$$\dot{\mathbf{x}} = \mathbf{F}(\mathbf{x}) + \zeta \tag{4.1}$$

where $\mathbf{x} = \{x_1(t), x_2(t), \ldots, x_n(t)\}$ is the concentration vector, each component of which represents different protein species in the network. The $\mathbf{F}(\mathbf{x}) = \{F_1(\mathbf{x}), F_2(\mathbf{x}), \ldots, F_n(\mathbf{x})\}$ is the chemical reaction rate flux vector involving the chemical reactions which are often nonlinear in protein concentrations $\mathbf{x}$ (for example, enzymatic reactions). The equations $\dot{\mathbf{x}} = \mathbf{F}(\mathbf{x})$ describe the averaged dynamical evolution of the chemical reaction network (see details in the next subsection). As mentioned, in the cell, the fluctuations can be very significant from both internal and external sources [28–33] and in general can not be ignored. A term ζ is added then as the noise mimicking these fluctuations in an assumed Gaussian distribution (from the large-number theorem in statistics). Then the auto correlation of the noise is given by:

$$< \zeta(\mathbf{x}, t)\zeta^\tau(\mathbf{x}', t') >= 2D(\mathbf{x}, t)\delta(t - t'). \tag{4.2}$$

Here $\delta(t)$ is the Dirac delta function and the diffusion matrix D is explicitly defined by $< \zeta_i(t)\zeta_j(t') >= 2D_{ij}\delta(t - t')$. The average $< \cdots >$ is carried out with the Gaussian distribution for the noise.

There exists a transformation [12], such that we can write the network equations into the following form:

$$(\mathbf{S}(\mathbf{x}) + \mathbf{A}(\mathbf{x}))\dot{\mathbf{x}} = -\partial U(\mathbf{x}) + \xi, \tag{4.3}$$

with the semi-positive definite symmetric matrix function $\mathbf{S}(\mathbf{x})$, the antisymmetric matrix $\mathbf{A}(\mathbf{x})$, the single-valued scalar function $U(\mathbf{x})$, and the stochastic force ξ. Here ∂ is the gradient operator in the state variable space. It is important to realize that the semi-positive definite symmetric matrix term is "dissipative": $\dot{\mathbf{x}}^T S(\mathbf{x})\dot{\mathbf{x}} >= 0$ ($\mathbf{x}^T$ is the transpose vector of $\mathbf{x}$); the antisymmetric part does no "work": $\dot{\mathbf{x}}^T A(\mathbf{x})\dot{\mathbf{x}} = 0$, therefore, is nondissipative. Hence, it is natural to identify that the dissipation is represented by the semi-positive definite symmetric matrix $\mathbf{S}(\mathbf{x})$, the friction matrix, and the transverse force by the antisymmetric matrix $\mathbf{A}(\mathbf{x})$, the transverse matrix. In general, we can not write $\mathbf{F}(\mathbf{x})$ as a gradient of a potential energy function and therefore no potential energy can be defined in noise-free environments. However, through the transformation, there exists a gradient term involving U in the presence of noise. The scalar function $U(\mathbf{x})$ then acquires the meaning of potential energy.

The steady-state distribution function $P_0(\mathbf{x})$ for the state variable $\mathbf{x}$ (representing the protein concentrations of the MAP-kinase signal transduction network in this case) can be shown to be exponential in potential energy function $U(\mathbf{x})$ [12, 16–18, 20, 21, 42]:

$$P_0(\mathbf{x}) = \frac{1}{Z} \exp\{-U(\mathbf{x})\}, \tag{4.4}$$

with the partition function $Z = \int d^d x \, \exp\{-U(x)\}$. From the steady-state distribution function, we can therefore identify U as the generalized potential energy function of the network system. In this way, we map out the potential landscape. Once we have the potential energy landscape, we can discuss the global stability of the protein cellular networks. Below are the detail descriptions of the calculation procedures.

Average kinetics

Here, we can start with a quantitative computational description of the MAPK cascade for average kinetics [2, 3]. Figure 4.1 illustrates the reaction schemes of the network. In step 1, the activation of MKKK is catalyzed by an enzyme E1; in step 2, the reverse is catalyzed by an enzyme E2. Steps 3 and 4 represent the two-collision, nonprocessive mechanism for the double phosphorylation of MKK; these steps are catalyzed by MKKK-P while their reverses are catalyzed by a phosphatase KKPase. Likewise, steps 7 and 8 represent the double phosphorylation of MAPK catalyzed by the active MKK-PP, and the reverse steps are catalyzed by a phosphatase KPase.

Based on the Michaelis–Menten enzyme kinetic equation, one can derive a set of differential equations which describe the variation rate of each component's concentration in the cascade [2, 3]. The associated chemical reaction rate coefficients are measured or inferred from the experiments in *Xenopus* oocytes [2, 3]. Together with the conservation equations, we have five independent simplified equations:

$$\frac{dx_2}{dt} = \frac{V_1(100 - x_2)}{K_1 + (100 - x_2)} - \frac{V_2 x_2}{K_2 + x_2} = F_2 \quad V_1 = 2.5, K_1 = 10, V_2 = 0.25, K_2 = 8 \tag{4.5}$$

$$\frac{dx_3}{dt} = \frac{V_6(300 - x_3 - x_5)}{K_6 + 300 - x_3 - x_5} - \frac{k_3 x_2 x_3}{K_3 + x_3} = F_3 \quad k_3 = 0.025, K_3 = 15, V_6 = 0.75, K_6 = 15 \tag{4.6}$$

$$\frac{dx_5}{dt} = \frac{k_4 x_2(300 - x_3 - x_5)}{K_4 + 300 - x_3 - x_5} - \frac{V_5 x_5}{K_5 + x_5} = F_5 \quad k_4 = 0.025, K_4 = 15, V_5 = 0.75, K_5 = 15 \tag{4.7}$$

$$\frac{dx_6}{dt} = \frac{V_{10}(300 - x_6 - x_8)}{K_{10} + 300 - x_6 - x_8} - \frac{k_7 x_5 x_6}{K_7 + x_6} = F_6 \quad k_7 = 0.025, K_7 = 15, V_{10} = 0.5, K_{10} = 15$$

(4.8)

$$\frac{dx_8}{dt} = \frac{k_8 x_5(300 - x_6 - x_8)}{K_8 + 300 - x_6 - x_8} - \frac{V_9 x_8}{K_9 + x_8} = F_8 \quad k_8 = 0.025, K_8 = 15, V_9 = 0.5, K_9 = 15$$

(4.9)

where x's are the concentrations of the corresponding proteins:

$$x_1 = [\text{MKKK}], x_2 = [\text{MKKK} - \text{P}], x_3 = [\text{MKK}], x_4 = [\text{MKK} - \text{P}],$$

$$x_5 = [\text{MKK} - \text{PP}], x_6 = [\text{MAPK}], x_7 = [\text{MAPK} - \text{P}], x_8 = [\text{MAPK} - \text{PP}]$$

and the concentrations (nm) are conserved respectively [2, 3]:

$$[\text{MKKK}]_{\text{total}} = x_1 + x_2 = 100;$$

$$[\text{MKK}]_{\text{total}} = x_3 + x_4 + x_5 = 300;$$

$$[\text{MAPK}]_{\text{total}} = x_6 + x_7 + x_8 = 300.$$

Potential energy landscape in fluctuating environments

Due to the intrinsic statistical fluctuations of the protein (kinase) numbers in the limited cell volume and external fluctuations within the cellular environments, the average descriptions of the chemical rate equations above need to be modified. We add noise sources to each rate equation and derive five stochastic differential equations:

$$\frac{dx_i}{dt} = F_i + \xi_i \quad (i = 2, 3, 5, 6, 8)$$

(4.10)

For simplicity we will assume that the noise is Gaussian and white with variance; it is equivalent that the means of the noise terms are zero: $< \xi_i(t) > = 0$ and the variance satisfies: $< \xi_i(t)\xi_j(t') > = 2D_{ij}\delta(t - t')$. Here D_{ij} is the diffusion matrix. As mentioned, there exists a transformation [12, 17, 18] such that we can transform the network equations into the form of equation (4.3), where the longitudinal frictional force term S, the transverse force term A, and potential function U for the network system are given as:

$$\begin{cases} U(\mathbf{x}) = - \int_C d\mathbf{x}' G^{-1}(\mathbf{x}') F(\mathbf{x}') \\ S(\mathbf{x}) = [G^{-1}\mathbf{x} + (G^t)^{-1}\mathbf{x}]/2 \\ A(\mathbf{x}) = [G^{-1}\mathbf{x} - (G^t)^{-1}\mathbf{x}]/2 \end{cases}$$

(4.11)

G and its transpose G^t satisfy constraint equations (G^{-1} is the inverse function of G):

$$G + G^t = 2D.$$

(4.12)

and

$$GV'_{\mathbf{x}} * [G^{-1}(\mathbf{x}')|_{\mathbf{x}'->\mathbf{x}}F(\mathbf{x})]G^t + GS^t - SG^t = 0. \tag{4.13}$$

One can approximate the above equation in gradient expansions to zero, first and high orders to solve for G and substitute the solution of G to obtain the potential energy U from (4.11). We for simplicity only solve the G and the corresponding U up to zero and first order. We found convergent solutions.

The zero-th order approximation of the G is given below:

$$GS^t - SG^t = 0, \quad S_{ij} = \frac{\partial F_i}{\partial x_j}. \tag{4.14}$$

We assume D is a diagonal matrix $D_0 I$ (I is the identity matrix) for simplicity, further, we take D_0 as constant 1. Then for the first-order solution, we can solve the linear set of equations for G. Taking G and performing the calculation of the integral, we can obtain the potential energy U.

Numerical solution

Equation (4.12) is a set of linear equations which can determine $5 * (5 + 1)/2$ conditions. Equation (4.13) can determine $5 * (5 - 1)/2$ conditions.

For a higher-order approximation, we have an iteration equation which takes the zero order result as an initial value. This equation reads:

$$G_{l-1}\nabla_{q'} \times [G^{-1}_{l-1}(q')|_{q'\to q}f(q)]G^t_{l-1} + G_l S^t - SG^t_l = 0. \tag{4.15}$$

This equation also determines $5 * (5 - 1)/2$ conditions. After we have G and G^{-1}, we can easily integrate (4.11) to obtain the potential energy U.

In the process of solving the linear set of equations for G, we first normalized the concentrations and then divided them into 20–1000 bins. We solved the problem with fewer bins and data points, but used a Monte Carlo method to sample the data with more bins and data points. We solved the value of G up to zero and first order and found the convergent solution [17].

Results and discussions

The potential energy U is a multidimensional function in concentration vector $\mathbf{x}$ space, each component of which represents the concentration of a different type of proteins. For certain configurations of concentration distribution, the network adopts certain potential energy (or the corresponding probability).

The configurational state space is huge. First of all, we are interested in which out of all the possible configurations is the most probable one or has the lowest energy state. We found the lowest energy state or the most probable configuration is the one at the end stage (ground state) of the MAP-kinase signal transduction, which corresponds to the fixed point steady-state solution

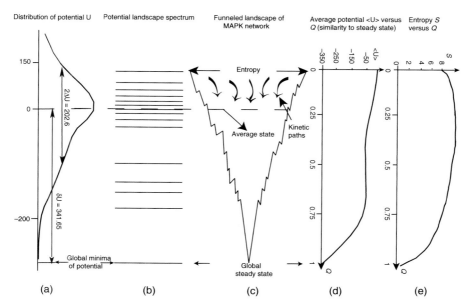

Fig. 4.2 The global structures and properties of the underlying potential landscape of the MAP-kinase signal transduction network. (a) The histogram or the distribution of the potential energy U. (b) The potential landscape spectrum. (c) The funneled landscape of the MAP-kinase signal transduction network. (d) The averaged potential energy $< U >$ as a function of similarity parameter Q with respect to the global minimum (or global steady-state fixed point). (e) The configurational entropy S_0 of the MAP-kinase signal transduction network as a function of similarity order parameter Q with respect to the global minimum (or global steady-state).

of the averaged chemical rate equations for the MAP-kinase network without the noise (Fig. 4.2a, b). This state is only one configurational state out of a vast number of possible ones, but it has the highest probability of occurring. The other configurational states have much lower probabilities of occurring.

Figure 4.2a shows a histogram or distribution of U. We can see that the distribution is approximately Gaussian. The lowest potential energy U is the global minimum of the potential energy landscape. Of course we can not visualize the potential energy function $U(\mathbf{x})$ in multidimensional configurational space of protein concentrations around this ground state. So we do a low-dimensional projection and look at the nature of the underlying potential landscape U. We did a zero-dimensional projection by just registering the energy spectrum which is illustrated in Fig. 4.2b. It is clear that the global minimum of the potential energy is significantly separated from the rest of the potential energy spectrum or distribution.

To quantify this, we define the robustness ratio RR for the network as the ratio of the gap δU (the difference between this global minimum $U_{global-minimum}$ and the average of U, $< U >$) versus the spread or half-width of the distribution of U, ΔU. Thus $RR = \delta U / \Delta U$. The δU is a measure of the bias or the slope towards the global minimum of the potential landscape, while ΔU is a measure of the averaged roughness or the local trapping of the potential landscape. When RR is significantly larger than 1, the gap is significantly larger than the roughness or local trapping of the underlying landscape. Then the global minimum is well separated and distinct from the rest of the network potential energy spectrum. Since $P_0(x) = 1/Z \exp\{-U(x)\}$, the weight or population of the global minimum will be the dominant one with a large RR. The populations of the rest of the possible states are much less significant. This leads to the global stability or robustness discriminating against others. The RR value for the MAP-kinase network is found to be $RR = 3$, significantly larger than 1. So RR gives a quantitative measure of the statistical properties characterizing the global topography of the underlying landscape. Only a cellular network landscape with a large RR will be able to form a stable global steady-state and be robust, thus surviving the natural evolution.

Figure 4.2d shows the one-dimensional projection of the averaged U, $< U >$, to the overlapping order parameter Q with respect to the global minimum ($Q = \sum_i^N x_i x_i^{global} / |\mathbf{x}||\mathbf{x}^{global}|$). The Q is defined this way so that we can keep track of the degree of "closeness" or overlap between an arbitrary state $\mathbf{x}$ to the global minimum state $\mathbf{x}_{global}$ in the state space of the protein concentrations. $Q = 1$ represents the global minimum state and $Q = 0$ represents the states with no overlap (decorrelated) to the global minimum. We see a downhill slope of the potential energy $< U >$ in Q towards the global minimum U_{global}. This shows clearly a funnel of $< U >$ along Q towards the global minimum of the potential landscape. When the chemical rate coefficients randomly change ($10 - 15\%$), the slope of $< U >$ changes (Fig. 4.3). The random changes of the rate coefficients correspond to smaller or shallower slope of U in Q towards the global minimum. This means that the landscape is less funneled. But the degree of the change for the slope of U in Q is mild, so the landscape is still biased towards the global minimum under a variety of quite different cellular conditions (Fig. 4.3). Therefore the network is stable and quite robust. With more drastic changes to the rate parameters (up to $20 - 50\%$), the global minimum starts to deviate from the steady state. The steepness of the funnel is less and the landscape becomes more flat (Fig. 4.3). This implies that the possible parameter window around experimental observed or inferred values of the chemical rate coefficients is finite. This limited range of parameters for the chemical rate coefficients leads to a funneled underlying potential landscape. The

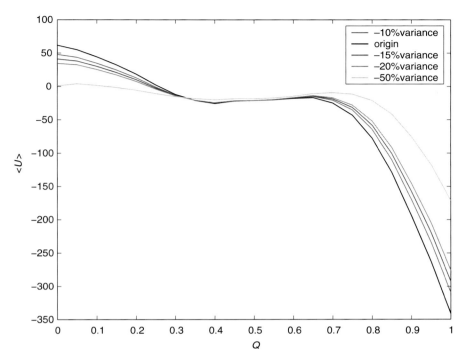

Fig. 4.3 Potential landscape funnel as a function of overlap order parameter Q against random changes of rate coefficients.

evolution selection will prefer a funneled landscape to perform the biological task of unidirectional and efficient signal transduction. The network is quite robust in this parameter range as shown above. On the other hand, the other part of the parameter space is huge (corresponding to even more drastic change around the experimentally observed or inferred rate coefficients) and unlikely to produce a funnel. The network in this parameter range therefore will not be preferred by the natural evolution selection and is more likely to be extinct.

Figure 4.2e shows the configurational entropy $S_0(Q)$ as a function of Q. The entropy is calculated by dividing the concentration variables into a multidimensional lattice, counting the number of states in each multidimensional lattice cube, and then is projected on Q. As we can see the entropy is rather smooth at small Q and decays as Q migrates towards the global steady state or global minimum. Since the entropy represents the number of states available, this implies the configurational state space for the network becomes smaller towards the global steady state. Thus entropy can be used as a measure of the radius of the funneled landscape perpendicular to the direction of the funnel towards the global steady state (see Fig. 4.2c that the funnel in radial size shrinks towards the global steady state).

We constructed the effective free energy versus overlap Q, $F(Q)$, by making use of the micro-canonical ensemble: $F(Q) = U - TS$. U and S are the potential energy and entropy of the system respectively. They are given by $U = <U>$ $(Q) - \Delta U^2(Q)/T$ and $S = S_0(Q) - \Delta U^2(Q)/2T^2$ [17, 18, 41]. Here the $<U>(Q)$ is the average of the potential energy of U at each overlap Q, the T is the fictitious temperature mimicking the strength of the fluctuations, the $\Delta U^2(Q)$ is the variance of the potential energy at each Q, and the $S_0(Q)$ is the entropy of the configuration at Q, which is given by $S_0(Q) = \ln \Omega(Q)$. The $\Omega(Q)$ is the number of the configurational states at particular overlap Q.

Figure 4.4 shows the free-energy profile of the network at different temperatures. At low temperatures, the free energy is biased more towards the global minimum of U and $Q = 1$ state is thermodynamically more stable; at high temperatures, the free energy is biased more towards the states that are less correlated with the global minimum of U and $Q < 1$ states are more thermodynamically stable. At intermediate temperature regime, the expression for free energy can have a double minimum in order parameter Q.

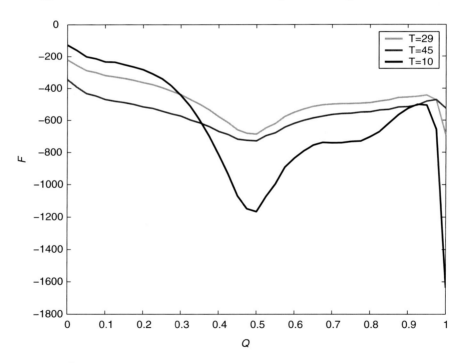

Fig. 4.4 Free energy as a function of overlap order parameter Q at high, intermediate, and low temperatures. At low temperatures, native phase with global minimum or steady-state fixed point is more preferred; at intermediate temperatures, less overlapped and native states are both preferred; at high temperatures, less overlapped states to the native one are more preferred.

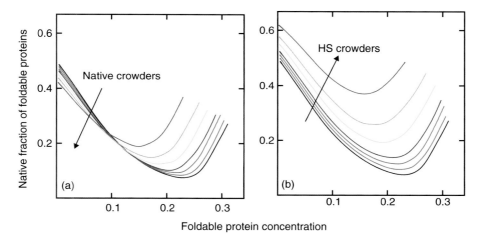

Plate 1 Native fraction of foldable proteins as a function of foldable protein concentration for the protein $N_r = 154$, $\Phi = 0.455$ at its infinite dilution folding temperature. Protein concentration is defined as in Fig. 1.4. The arrow indicates increasing concentration of (a) ultrastable native crowders and (b) inert hard-sphere (HS) crowders.

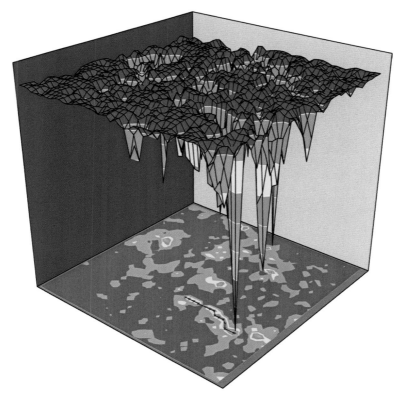

Plate 2 The potential landscape of the yeast cell cycle network and biological path to stationary G1. The lowest energy state corresponds to stationary G1 state. The green band with arrows corresponds to the biological path (sequentially from state 1 to 13 described in the text).

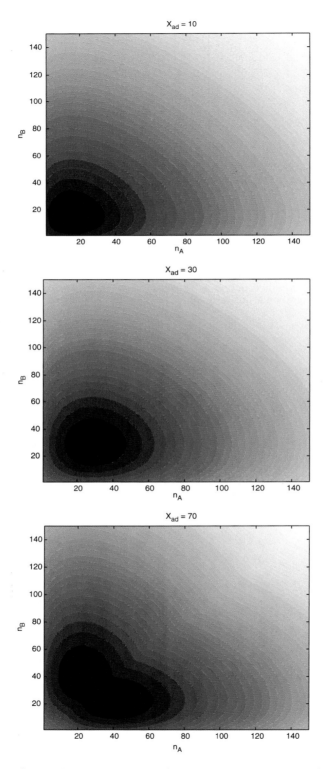

Plate 3 The potential energy of symmetric toggle switch as a function of the numbers of protein A and the numbers of protein B for different $X_{ad} = g_{A1}/2k_A$. The other parameters are the same as Fig. 4.15.

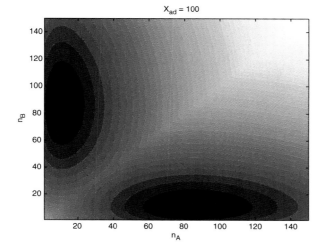

Plate 3 (Cont.)

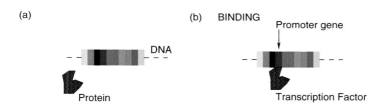

(a)

DNA

Protein

(b) BINDING

Promoter gene

Transcription Factor

(c) TRANSCRIPTION

mRNA

Activated gene

(d) TRANSLATION

Ribosome

Protein product

Rep

Act

Other

Plate 4 The different stages of gene expression. (a) The basic ingredients are the proteins and specific gene regions in the DNA, such as promoters and transcribed sequences. (b) A specific protein binds to a part of the DNA sequence called the promoter; the protein is known as the transcription factor since it starts the transcription of the genetic information encoded at the specific gene that the complex promoter + transcription factor regulates. (c) After the genetic information is transcribed into the messenger RNA, by RNA polymerase, it is subsequently translated into proteins at the ribosomes (panel d). The protein product that emerges after this process can act either as another transcription factor for the expression of other genes or as a repressor of the activity of other genes stopping the synthesis of their protein products. Another possibility is that this protein product participates in the physiological processes of the cell and forms protein complexes such as enzymes.

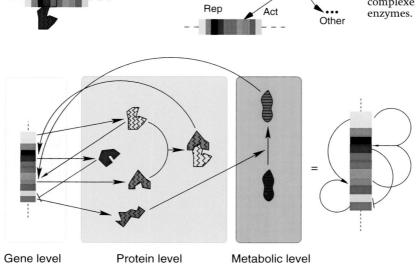

Gene level   Protein level   Metabolic level

Plate 5 Coarse-graining of cellular interactions into a single gene network. The three levels of description (genes, proteins, and metabolism) and the interactions between their constituents are embedded on a single map of interactions between genes.

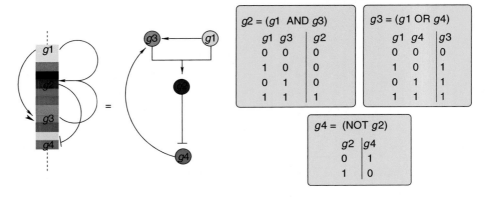

Truth tables

g2 = (g1 AND g3)		
g1	g3	g2
0	0	0
1	0	0
0	1	0
1	1	1

g3 = (g1 OR g4)		
g1	g4	g3
0	0	0
1	0	1
0	1	1
1	1	1

g4 = (NOT g2)	
g2	g4
0	1
1	0

Plate 6 Translation of the regulatory genetic map of Fig. 6.6 into Boolean regulatory functions. The three Boolean relations for genes g_2, g_3 and g_4 make use of the basic logical operators AND, OR, and NOT, respectively.

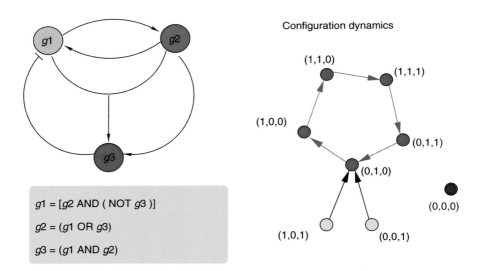

Configuration dynamics

$g1 = [g2$ AND (NOT $g3$)]

$g2 = (g1$ OR $g3)$

$g3 = (g1$ AND $g2)$

Plate 7 Small regulatory network and its configuration dynamics. The network is composed of three genes that interact following the logical rules shown in the box. The configuration dynamics is represented by a network whose nodes are all the possible dynamical states and the directed links are the transitions from one to another (as dictated by the Boolean dynamics). It is shown that a cycle of length 5 (red nodes) and a steady state (blue node) exist. Yellow nodes are in the basin of attraction of the periodic attractor.

There exist two characteristic thermodynamic transition temperatures for the network system. We can equate the two minimum of the free energies from less-overlapping states to the global minimum state of the MAP-kinase network to obtain the native transition temperature to the global minimum $(F(Q = Q^*) = F(Q = 1))$:

$$T_n = \frac{<U>(Q=Q^*)- <U>(Q=1)}{2S_0(Q=Q^*)}\left(1 + \sqrt{1 - \frac{2S_0(Q=Q^*)\Delta U^2(Q=Q^*)}{(<U>(Q=Q^*)- <U>(Q=1))^2}}\right).$$

$$(4.16)$$

Here $<U>(Q=Q^*) - <U>(Q=1)$ is the gap between the global minimum and the less-overlapped states (the states which have the same free energy as the global minimum). There also exists another possible transition where the entropy of the system goes to zero. This means that the system runs out of the states and gets trapped into a local minimum. This gives $S_0(Q=Q^*) = \Delta U^2(Q=Q^*)/2T^2$ so that the trapping temperature is given by $T_{\text{trapping}} = \sqrt{\Delta U^2(Q=Q^*)/2S_0(Q=Q^*)}$.

Taking the ratio of native transition temperature and trapping temperature of the network, we obtain:

$$T_n/T_{\text{trapping}}(Q = Q^*) = \Lambda + \sqrt{\Lambda^2 - 1}$$

$$(4.17)$$

where $\Lambda = [< U(Q = Q^*) > - < U(Q = 1) >]/[\Delta U(Q = Q^*)\sqrt{2S_0(Q = Q^*)}]$ is the ratio of the potential energy gap between the global minimum state and the average of the potential energy landscape spectrum versus the ruggedness or the half-width (spread) of the distribution of the potential energy landscape spectrum, weighted by entropy or the measure of the number of the states available $\sqrt{2S_0(Q = Q^*)}$. We can see RR is directly related to Λ: $\Lambda = RR/\sqrt{2S_0(Q = Q^*)}$.

There are at least three possible thermodynamic phases: the global minimum, the less overlapping with the global minimum, and trapping phase (see Fig. 4.5). The global minimum state in the MAP-kinase example corresponds to the final destination at the end of the signal transduction network. Clearly, the temperature of native transition to the global minimum should be higher than the trapping temperature in order to guarantee the global thermodynamic stability and avoid nondiscrimination with traps. In order to assure that, the ratio T_n/T_{trapping} should be maximized. From the above expression, this is equivalent to saying that Λ (or robustness ratio RR weighted by entropy) should also be maximized.

Therefore maximizing the ratio of the potential energy gap (or the slope) versus the roughness of the underlying potential energy landscape weighted by the

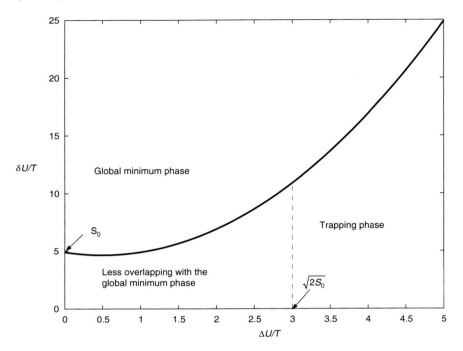

Fig. 4.5 Thermodynamic phase diagram for the MAP-kinase signal transduction network. Native phase with global minimum or steady-state fixed point; nonnative phase with states less overlapping with global minimum or steady-state fixed point; trapping phase with states trapped into the local minimum. The larger $\delta U/T$ and smaller $\Delta U/T$, or the larger $\delta U/\Delta U$, the more likely the global minimum state is thermodynamically stable and robust.

entropy of the available states (a measure of the configurational search space) becomes the criterion for the global thermodynamic stability or robustness of the network. Only the cellular network landscape satisfying this criterion will be able to form a thermodynamically stable global steady state, be robust, perform the biological functions, and furthermore survive the natural evolution. As in the case of the protein folding and binding problems [40, 41], this implies a funneled potential landscape of cellular network as shown in Fig. 4.2c, which has a directed downhill slope biasing towards the global minimum state, dominating the fluctuations or wiggles superimposed on the landscape and the configurational search space. From this picture, at the initial stage of the signal transduction network process, there are multiple parallel paths leading towards the global minimum state. As the kinetic process progresses, discrete paths might emerge and give dominant contributions when the roughness of the underlying landscape becomes significant.

Budding yeast cell cycle

To explore the nature of the underlying potential landscape of the cellular network, we will study the budding yeast cell cycle network [18, 21]. One of the most important functions of the cell is reproduction and growth. It is therefore crucial to understand the cell cycle and its underlying process. The cell cycles during development are usually divided in several phases: G1 phase in which the cell starts to grow under appropriate conditions, S phase in which DNA synthesis and chromosome replication occurs, G2 phase where the cell is in the stage of preparation for mitosis, and M phase in which chromosome separation and cell division occurs. After passing through the M phase, the cell enters back to G1 phase and thus completes a cell cycle. In most eukaryotic cells, the elaborate control mechanisms over DNA synthesis and mitosis make sure the crucial events in the cell cycle are carried out properly and precisely. Physiologically, there are usually several check points (where cells are in the quiescent phase waiting for the signal and suitable conditions for further progress in the cell cycle) for controlling and coordination: G1 before the new round of division, G2 before the mitotic process begins, and M before segregation.

Recently, many of the underlying controlling mechanisms are revealed by the genetic techniques such as mutations or gene knockouts. It is found that control has been centered around cyclin-dependent protein kinases (CDKs) which trigger the major events of the eukaryotic cell cycle. For example, the activation of cyclin/CDK dimer drives the cells at both G1 and G2 checkpoints for further progress. During other phases, CDK/cyclin checkpoints are activated. Although molecular interactions regulating the CDK activities are known, the mechanisms of the checkpoint controls are still uncertain [34–37].

The cell-cycle process has been studied in details in the budding yeast *Saccharomyces cerevisiae* [4, 11, 34–37]. There are many genes involved in controlling the cell cycle processes. But the number of the crucial regulators is much less. A network wiring diagram based on the crucial regulators can be constructed [11, 34–37] as shown in Fig. 4.6.

Under the rich nutrient conditions and when the cell size grows large enough, a cyclin Cln3 will be turned on. Thus the cell-cycle sequence starts when the cell commits to division through the activation of Cln3 (the START). Cln3/Cdc28 will be activated. This in turn activates through phosphorylation a pair of transcription factor groups, SBF and MBF, which activate the genes of the cyclins Cln1 and Cln2 and Clb5 and Clb6, respectively. The subsequent activity of Clb5 drives the cell into the S phase where DNA replication begins. The entry into the M phase for segregation is controlled by the activation of Clb2 through the transcription factor MCM1/SFF activation. The exit from the M phase is controlled

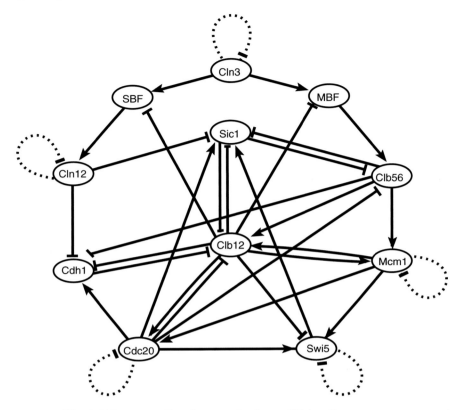

Fig. 4.6 The yeast cell cycle network scheme. Wiring diagram: arrow represents positive activating regulations (1); arrow ⊣ represents negative suppressing regulations (−1); arrow ··⊣ represents self-degradation.

by the inhibition and degradation of Clb2 through the Sic1, Cdh1, and Cdc20. Clb2 phosphorylates Swi5 to prevent its entry into the nucleus. After the M phase, the cell comes back to the stationary G1 phase, waiting for the signal for another round of division. Thus the cell-cycle process starts with the excitation from the stationary G1 state by the cell-size signal and evolves back to the stationary G1 state through a well-defined sequence of states.

Mathematical models of the cell cycle controls have been formulated with a set of ordinary first-order (in time) differential equations mimicking the underlined biochemical processes [11, 34–37]. The models have been applied to the budding yeast cycle and explained many qualitative physiological behaviors. The checkpoints can be viewed as the steady states or stationary fixed points. Since the intracellular and intercellular signals are transduced into the changes in the regulatory networks, the cell cycle becomes the dynamics in and out of the fixed points. Although detailed simulations give some insights towards the issues, due to the limitation of the parameter space search, it is difficult to perceive the global or universal properties of the cycle networks (for example, for different species). It is the purpose of the current study to address this issue.

We will study the global stability by exploring the underlying potential landscape for the yeast cell cycle network [18, 21].

Methods

The average dynamics of the network can be usually described by a set of chemical rate equations for concentrations where both the concentrations and the links among them through binding rates with typically quite different timescales are treated in a continuous fashion. In the cell cycle, most of the biological functions seem to arise from the on and off properties of the network components. Furthermore, the global properties of the network might depend less sensitively on the details of the model. Therefore, a simplified representation [11] can be proposed with each node i having only two states $S_i = 1$ and $S_i = 0$, representing the active and the inactive state of the protein, or high concentration and low concentration of proteins, respectively. As illustrated in Fig. 4.6, we have 11 protein nodes in the network wiring diagram, we have all together 2^{11} states, each state represented by S with a distinct combination of the on and off of the 11 protein nodes of Cln3, MBF, SBF, Cln1-2, Cdh1, Swi5, Cdc20, Clb5-6, Sic1, Clb1-2, Mcm1 represented by $\{S_1, S_2, S_3, ..., S_{11}\} = S$. We can then define some rules to follow the subsequent dynamics of the network. Therefore the evolution of the network is deterministic.

As mentioned, in the cell the average dynamics of the cellular network might not give a good description of the system. This is due to the intrinsic fluctuations from the limited number of the proteins in the cell and extrinsic fluctuations from the environments in the interior of the cell. It is then more appropriate to approach the network dynamics based on statistical description. In other words, we should replace the deterministic or average description of the dynamics of states in the cellular network to a probabilistic description of the evolution of the cellular network dynamics. So instead of following the on and off switching in the network, we follow the probability of on and off for each state in the network.

In order to follow the evolution of the states in the cellular network, we need first to figure out the transition probability from one state S_1 at present time to another state S_2 at the next moment. This is difficult to solve and in general almost impossible. We therefore will make some simplifications so that we can handle the case without loss of generality by assuming that the transition probability T from one state to another can be split into the product of the transition probability for each individual flip (or no flip) of the on or off state from this moment to the next moment. The transition probability from one state at current state to another at the next moment will be assumed not to depend on the earlier times (no memory). This leads to the Markovian process [43–45]. The transition matrix T can thus be written as:

$$T_{\{S_1(t'),S_2(t'),\dots,\,S_{11}(t')|S_1(t),S_2(t),\dots,\,S_{11}(t)\}} = \Pi_{i=1}^{11} T_{\{S_i(t')|S_1(t),S_2(t),\dots,\,S_{11}(t)\}} \tag{4.18}$$

where t is the current time and t' is the next moment. So the whole transition probability from current state to the next is split into the product of the transition probability of each individual flip (or no flip) of the node i. For each individual flip, the transition probability for a particular node can be modeled as a nonlinear switching function as shown in Fig. 4.7a and b from the input through the interactions to the output which is often used in neural science [46]:

$$T_{\{S_i(t')|S_1(t),S_2(t),\dots,S_{11}(t)\}} = \frac{1}{2} \pm \frac{1}{2} \tanh[\mu \sum_{j=1}^{11} a_{ij}S_j(t)]. \tag{4.19}$$

When the input $\sum_{j=1}^{11} a_{ij}S_j(t) > 0$ is positive (activation), the transition probability to the on state is higher (close to 1). When the input is negative (repression) the transition probability to the on state is lower (close to zero). Furthermore

$$T_{S_i(t')|S_1(t),S_2(t),\dots,S_{11}(t)} = 1 - c \tag{4.20}$$

when there is no input of activation or repression ($\sum_{j=1}^{11} a_{ij}S_j(t) = 0$); c is a small number mimicking the effect of self-degradation. Here a_{ij} is the arrow or link representing the activating $(+1)$ or suppressing (-1) interactions between ith and jth protein node in the network which is explicitly shown in the wiring diagram of Fig. 4.6, and μ is a parameter controlling the width of the switching function from the input to the output. The physical meaning is clear. If the inputs through the interactions among proteins to a specific protein node in the network are large enough, then the state will flip, otherwise the state will stay without the flip. The positive (negative) sign in the T expression gives the probability of flipping from 0(1) to 1(0) state. If μ is small (large), the transition width is large (small), the transition is smooth (sharp or sensitive) from the original state to the output state. Therefore we have an analytical expression of the transition probability (Fig. 4.7).

With the transition probability among different states specified, finally we can write down the master equation for each of the 2^{11} states as:

$$dP_i/dt = -\sum_j T_{ij}P_i + \sum_j T_{ji}P_j \tag{4.21}$$

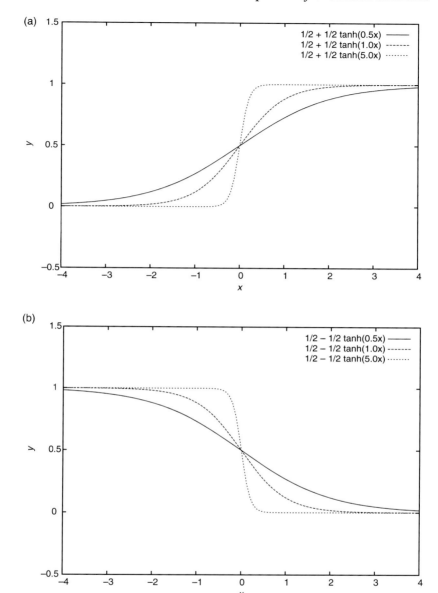

Fig. 4.7 Nonlinear response function versus inputs: (a) is for $y = 1/2 + 1/2 \tanh(\mu x)$ when $\mu = 0.5, 1, 5$ and (b) is for $y = 1/2 - 1/2 \tanh(\mu x)$ when $\mu = 0.5, 1, 5$.

where T_{ij} (T_{ji}) represents the transition probability from state $i(j)$ to state $j(i)$ specified in details above. Here i and j are from 1 to $2^{11} = 2048$ states and $\sum_{i=1}^{i=2^{11}} P_i = 1$.

We solved the $2^{11} = 2048$ master equations of the yeast cell cycle numerically (by using iterative method) to follow the evolution of the probability distribution of each state, with the initial condition of equal small probability of all the cell states ($P_i = 1/2048$). Both the time-dependent evolution and the steady-state probability distribution for each state are obtained.

Let us focus on the steady-state probability distribution. For each state, there is a probability associated with it. One can write the probability distribution for a particular state as $P_i = \exp[-U_i]$ ($\sum_{i=1}^{i=2^{11}} P_i = 1$) or $U_i = -\ln P_i$. One can immediately see that U_i acquires the meaning of generalized potential energy (from the Boltzman distribution). This is the key point: although there is no potential energy function directly from the normal deterministic averaged chemical reaction rate equations for the network, a generalized potential energy function does exist and can be constructed from the probabilistic description of the network instead of the deterministic averaged one. This generalized potential energy function is inversely related to the steady-state probability. When the probability is large, the potential energy is lower and when probability is small, the potential energy is higher. The dynamics of the cell-cycle thus can be visualized as passing through mountains and ridges of the potential landscape in state space of the cell-cycle network to the final destiny. The advantage of introducing the concept of energy is that once we have the potential landscape, we can discuss the global stability of the protein cellular networks. Otherwise, it is almost impossible to address the global issues without going through the parameter space locally which is often cosmologically big.

The network is an open system in nonequilibrium state. Even at steady state, the system is not necessarily in equilibrium. This is clear from the fact that although we can obtain the steady-state probability and can define an equilibrium like quantity such as steady-state probability, the flux is not necessarily equal to zero ($F_{ij\,\text{steady-state}} = -T_{ij}P_{i\,\text{steady-state}} + T_{ji}P_{j\,\text{steady-state}}$). This is different from the equilibrium situation where the local flux is equal to zero (detailed balance condition). The flux defines a generalized force for the nonequilibrium state along with the associated generalized chemical potential [47, 48]. The nonequilibrium state dissipates energy. In the steady state, the heat loss rate is equivalent to entropy production rate, where entropy S_0 is defined as $S_0 = -\sum_i P_i \ln P_i$ and entropy production rate (per unit time) $\dot{S}$ is given by:

$$\dot{S} = \sum_{ij} T_{ji}P_j \ln \left(\frac{T_{ji}P_j}{T_{ij}P_i} \right) \tag{4.22}$$

Entropy production rate is a characterization of the global properties of the network. We can study how the entropy production rate or dissipation cost of the network varies with the changes of internal and external perturbations. We can explore the global natures of the network such as stability, robustness, and dissipation cost and their interrelationships.

In each of the simulations, we study the robustness of the network by exploring different values of switching and self-degradation parameters μ and c, as well as the mutations of the links or interactions in the network [18, 21].

Results

Since the potential energy is a multidimensional function in protein states, it is difficult to visualize U. So we directly look at the energy spectrum (Fig. 4.8) and explore the nature of the underlying potential landscape U [18, 21].

Figure 4.8a shows the spectrum as well as the histogram of U. We can see that the distribution is approximately Gaussian. The lowest potential U is the global minimum of the potential landscape. It is important to notice this global minimum of U is found to be the same state as the steady state or fixed point (the stationary G1 state $= (0;0;0;0;1;0;0;0;1;0;0)$) of the deterministic averaged chemical reaction network equations for the yeast cell cycle. It is clear that the global minimum of the potential is significantly separated from the average of the potential spectrum or distribution.

To quantify this, we define the robustness ratio RR (similar to the MAPK case [17]) for the network as the ratio of the gap δU, the difference between this global minimum of G1 state $U_{\text{global-minimum}}$ and the average of U, $< U >$ versus the spread or the half-width of the distribution of U, ΔU, $RR = \delta U / \Delta U$ as shown in Fig. 4.8a; δU is a measure of the bias or the slope towards the global minimum (G1 state) of the potential landscape; ΔU is a measure of the averaged roughness or the local trapping of the potential landscape. When RR is significantly larger than 1, the gap is significantly larger than the roughness or local trapping of

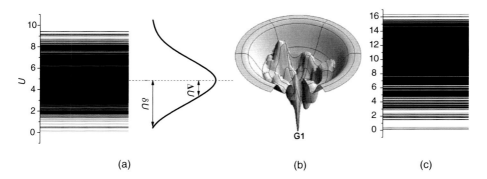

(a)   (b)   (c)

Fig. 4.8 The global structures and properties of the underlying potential landscape of the yeast cell cycle network. (a) The spectrum and the histogram or the distribution of the potential energy U. (b) An illustration of the funneled landscape of the yeast cell cycle network. The global minimum of the energy is at G1 state. (c) The spectrum of the potential energy U for a random network.

the underlying landscape, then the global minimum (G1 state) is well separated and distinct from the average of the network potential spectrum. Since $P = \exp\{-U(x)\}$, the weight or population of the global minimum (G1 state) will be dominated by the one with large RR. The populations of the other possible states are much less significant. This leads to the global stability or robustness discriminating against others. The RR value for the yeast cell cycle network is RR = 3 (for $\mu = 5$ and $c = 0.001$) as shown in Fig. 4.8a, significantly larger than 1. This shows a funnel picture of energy going downhill towards G1 state in the evolution of network states, as illustrated in Fig. 4.8b. So RR gives a quantitative measure of the property of the underlying landscape spectrum.

We found the typical values for random networks are close to 2 (RR can not be less than 1). A typical random network with RR $\approx$ 2 is illustrated in Fig. 4.8c for a random network. The ground state is not necessarily the G1 state any more. The probability of G1 is smaller for the random network compared with the biological one and therefore less stable. Thus, only the cellular network landscape with large value of RR will be able to form a stable global minimum G1 state, be robust, perform biological function and survive the natural evolution.

We identified the preferential global pathway towards the global minimum G1 by following the most probable trajectory in each step of the kinetic moves from the kinetic master equations towards G1. The protein can be either 1 or 0 representing active or inactive. The 11 proteins are arranged in a vector form to represent the state of the system as (Cln3; MBF; SBF; Cln1,2; Cdh1; Swi5; Cdc20; Clb5, 6; Sic1; Clb1,2; Mcm1). The most probable global path follows the states $1 - >13$ sequentially towards G1 from the start signal, where start signal is in state sequence 1 given by: (1;0;0;0;1;0;0;0;1;0;0). Three excited G1 states are in sequence 2, 3, 4, given respectively by: (0;1;1;0;1;0;0;0;1;0;0), (0;1;1;1;0;0;0;1;0;0), (0;1;1;1;0;0;0;0;0;0). The S phase is in state with sequence 5 given by: (0;1;1;1;0;0;0;1;0;0;0). The G2 phase is in state with sequence 6 given by (0;1;1;1;0;0;0;1;0;1;1). The M phase is in states with sequence 7, 8, 9, 10, 11, given respectively by: (0;0;0;1;0;0;1;1;0;1;1), (0;0;0;0;0;1;1;0;0;1;1), (0;0;0;0;0;1;1;0;1;1;1), (0;0;0;0;0;1;1;0;1;0;1), (0;0;0;0;1;1;1;0;1;0;0). The other excited G1 state is with sequence 12 given by (0;0;0;0;1;1;0;0;1;0;0). Finally stationary G1 phase is in state sequence 13 given by (0;0;0;0;1;0;0;0;1;0;0). The most probable path turns out to be the biological path going through $G1->S->G2->M->G1$.

We arranged the state space into the two-dimensional grids with the constraints of minimal overlapping or crossings of the state connectivity for clear visualization purpose. Each point on the two-dimensional grid represents a state (one of 2048 states). The potential landscape on the two-dimensional grids is

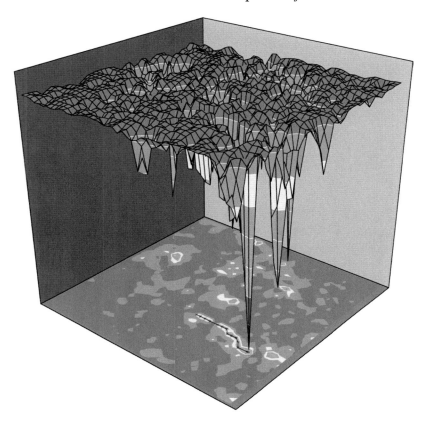

Fig. 4.9 The potential landscape of the yeast cell cycle network and biological path to stationary G1. The lowest energy state corresponds to stationary G1 state. The green band with arrows corresponds to the biological path (sequentially from state 1 to 13 described in the text). (See plate section for color version.)

shown in Fig. 4.9. The lowest energy state corresponds to the stationary G1 state. The global biological path is represented by the narrow green band on the projected two-dimensional state space plane. It is sequentially from state 1 to 13 as mentioned in the above text (sequences 1– >13). As we can see, the global biological path is in the low-energy valley of the landscape towards G1. In addition, we can also see some other off-pathway traps (states with low energies).

Figure 4.10a shows the robustness ratio (RR) of the cell cycle network versus the steady-state probability of the G1 (with $\mu = 5$ and $c = 0.001$) against various perturbations through deleting an interaction arrow, adding an activating or repressing arrow between the nodes that are not yet connected in the network wiring diagram in Fig. 4.6, or switching an activating arrow to a repressing

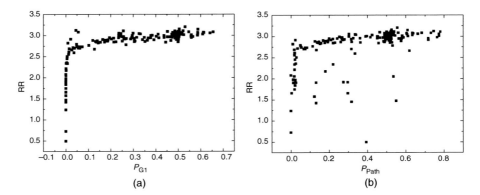

Fig. 4.10 Robustness against mutation perturbations. (a) Robustness ratio versus steady-state probability of G1, P_{G1} for different mutations of the links.
(b) Robustness ratio versus steady-state probability of biological path, P_{path}, for different mutations of the links.

arrow or vice versa, and deleting an individual node. There is a monotonic relationship between the G1 probability and RR. When RR is larger (smaller), the landscape is more (less) robust, the network is more (less) stable with G1 state dominating (less significant). Therefore RR is indeed a robustness measure for the network.

Figure 4.10b shows the robustness ratio (RR) versus steady-state probability of the global biological path with important biological states including G1 [11]. We see again that a network with large RR characterizing the funneled landscape leads to higher steady-state probability and therefore more stable biological path. Random networks typically have smaller RR and smaller probability of G1 compared with the biological one. They are less stable. The biological functioning network is quite different from the random ones in terms of the underlying potential landscape and stability.

Figure 4.11a shows the the robustness ratio of the underlying potential landscape versus different switching parameters μ ($c = 0.001$). We see that when μ is large (small) indicating a sharp (smooth) transition or response from input to output for a single flip of the protein states, the robustness ratio increases with μ increases. This means, a sharper transition or response from input to output gives a more robust network compared with the smoother transition or response. The value of μ can also be seen as a measure or characterization of the strengths of the noise from the intrinsic or extrinsic statistical fluctuations in the cellular environments [45]; μ could then be related to the inverse of the "temperature" (temperature here is a measure of the strength of the noise level). The energy U we defined in this paper is in units of μ. So U is a dimensionless

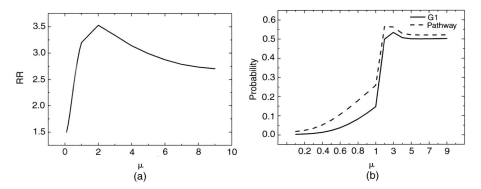

Fig. 4.11 Robustness against the sharpness of the response or the inverse noise level. (a) Robustness ratio versus sharpness of the response or inverse of noise level μ. (b) Steady-state probability of stationary G1, P_{G1}, and biological path, P_{path} versus μ.

quantity. When μ is not changing then the two definitions of U ($U = -\mu\log P$ and $U = -\log P$) are only different by a constant. The RR is not influenced by the above two definitions of U since it involves the ratio of the Us.

When μ is large, the transition is sharp. This corresponds to all-or-none deterministic behavior for the response or transition (0 or 1). This is the situation when the underlying statistical fluctuations are small. When μ is small, the response or the transition is no longer all-or-none (1 or 0) but a smooth function in between 0 and 1. This is due to the fact the statistical fluctuations lead to the states being more distributed and with less sharp responses. Therefore, the associated probability of distributed states has more chances being between 0 and 1. In other words, less (more) statistical fluctuations or sharper response with larger μ (less sensitive response with smaller μ) lead to a more (less) robust network characterized by large (small) RR. Then, there exist two phases for the network: a robust phase with RR significantly larger than 2, where the network is stable and the underlying potential landscape is funneled towards G1; and a fragile phase with RR dropping to around or below 2, where the network is less stable and the underlying potential landscape is more shallow towards G1.

Figure 4.11b shows the probability of stationary G1 state as well as the probability of the global path towards G1 versus μ. We can see a global transition phase transition at $\mu \sim 1$ below which P_{G1} and P_{path} significantly drops. From Fig. 4.11a and b, we see when P_{G1} and P_{path} is small, RR is also small, implying the system is less stable. Therefore the network loses the stability below $\mu \sim 1$. Significantly above $\mu \sim 1$, the network becomes stable. We can interpret this as the phase transition from the weak noise limit where the underlying landscape and the associated global path are not influenced much by the noise level to the limit where underlying landscape and associated global path are disturbed significantly or disrupted by the strong noise. We can also interpret this as the transition from hypersensitive response leading to the robustness of the landscape and the associated global path, to the inert or insensitive response leading to the fragile landscape and associated global path to G1. We can see a sharper response or more sensitivity of the individual protein nodes to the rest of the protein network through interactions usually leads to more robustness of the network with stable G1 and biological path.

The low μ corresponds to the strong noise limit or insensitive response for the node to the input. The landscape has low RR and is less stable or robust. The landscape is more flat and less biased towards G1. When μ increases, the noise level decreases, the response to the input is more sensitive for each node. This results in a more funneled towards G1 and more robust landscape. The maximal funnel is found around $\mu = 2$. There is a sharp change of the shape of the landscape near $\mu = 2$ from the $\mu < 2$ side. When $\mu > 2$, the RR value is slightly lower and quickly approaches to a constant as μ becomes larger corresponding to smaller noise, and more sensitive response from a node to the input. The landscape becomes stabilized with a definite RR and probability of P_{G1}. The peak value of the RR, P_{G1} as well as P_{path} implies there might exist traps in the landscape (deep energy states other than G1 and not on the biological path). Large noise will destroy the landscape which leads to low RR, P_{G1} as well as P_{path}. Zero noise leads to a relatively stable network with relatively large RR, P_{G1} as well as P_{path}. In the presence of traps, adding a small amount of noise helps the system to reach the global minimum without getting caught or trapped in the intermediate off-pathway trapping states. This increases the probability and enhances the stability of G1 and the biological path. Therefore the presence of the peak of RR, P_{G1} as well as P_{path} is an indication of the existence of traps in the landscape. We found six major off-pathway traps responsible for the peak in RR, P_{G1} as well as P_{path} (some are shown in Fig. 4.9).

Figure 4.12a shows the RR of the underlying energy landscape versus different self-degradation parameters c (at $\mu = 5$). We see that when c is large (small) indicating a large (small) self-degradation, the robustness ratio increases as c decreases. This means that less degradation gives a more robust network. Figure 4.12b shows the probability of stationary G1 as well as biological path versus different self-degradation parameters c (at $\mu = 5$). We see that when c is large (small) indicating a large (small) self-degradation, the probability of stationary G1 phase and biological path increases as c decreases. This means that less degradation gives a more probable and stable stationary G1 phase and biological path and therefore a more robust network.

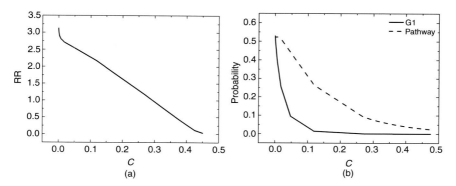

Fig. 4.12 Robustness against self-degradation. (a) Robustness ratio versus degree of self degradation, c. (b) Steady-state probability of stationary P_{G1} and biological path P_{path} versus c.

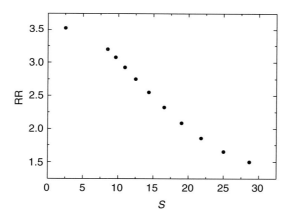

Fig. 4.13 Dissipation cost versus robustness of the network: entropy production rate S versus robustness ratio (RR).

In Fig. 4.13, we plotted the entropy production (per unit time) or the dissipation cost of the network, S, versus RR for different μ. We can see the entropy production rate decreases as RR increases. This implies the more robust the network is, the less entropy production or heat loss in the network. This can be very important for network design. Nature might evolve such that the network is robust against internal (intrinsic) and environmental perturbations, and perform specific biological functions with minimum dissipation cost. The fact that robustness is linked with the entropy production rate may reflect the fact that less fluctuations and perturbations leads to more robust and stable network, also more energy saving, and therefore lower costs in the meantime.

This might provide us a design principle of optimizing the connections of the network with minimum dissipation cost for the network. In our study here this is also equivalent to optimizing the robustness or stability of the network.

Discussions

The potential landscape is a statistical-based approach which is good in two respects. It is an approach that captures the global properties. On the other hand, the statistical approach can be very useful and informative when the data are rapidly accumulating. In this picture, there are many possible energy states of the network corresponding to different patterns of combinations of activation and inhibition of the protein states. Each checkpoint can be viewed as basins of attractions of globally low-energy states. The G1 phase state has the lowest global energy since it is the end of the cycle. We think it might be possible to describe the cell cycle as the dynamic motion in the potential landscape state space from one basin to another. This kinetic search can not be entirely random but directed since the random search takes cosmological time. The direction or gradient of the landscape is provided from the bias in terms of the energy gap towards the G1 phase. So the landscape picture becomes that there is a funnel towards the G1 state (the bottom of the funnel, that we can call the native state). At the end of G1 phase, the network is primed upon receiving the new start signal or nutrition (without those, the system will stay at G1 and the network can not continue the cycling process) to high-energy excited states at the top of the funnel (cycling). Then the cell cycle follows as it cascades through the configurational state space (or potential landscape) in a directed way passing several checkpoints (basins of attraction) and finally reaches the bottom of the funnel G1 phase before being pumped again for another cycle (Fig. 4.9).

We can see from the above discussions that maximizing the ratio of the potential gap (or the slope) versus the roughness of the underlying potential landscape is the criterion for the global stability or robustness of the network. Only the cellular network landscape satisfying this criterion will be able to form a stable global steady state, be robust (Figs. 4.10–4.12), perform the biological functions with minimal dissipation cost (Fig.4.8) and survive the nature evolution. Similar to protein folding and binding problems [40, 41], a funneled potential landscape of cellular networks emerges. The landscape biases towards the global minimum G1 state and dominates the fluctuations or wiggles in the configurational space. From this picture, at the initial stage of the yeast cell cycle network process, there could be multiple parallel paths leading towards the global minimum G1 state. As the kinetic process progresses, the discrete paths might emerge and give dominant contributions (biological path) when the roughness of the underlying landscape becomes significant (Fig. 4.9).

Cellular networks with too rough an underlying potential landscape can neither guarantee the global robustness nor perform specific biological function. They are more likely to phase out from evolution. The funneled landscape therefore is a realization of the Darwinian principle of natural selection at the cellular network level. As we see, the funneled landscape provides an optimal criterion for selecting the suitable parameter subspace of cellular networks, guaranteeing the robustness and performing specific biological function with less dissipation cost. This will lead to an optimal way for the network connections and is potentially useful for the network design.

It is worth pointing out that the approach described here is general and can be applied to many cellular networks such as signaling transduction network [2, 17], metabolic network [49], and gene regulatory network [10, 13, 20].

Toggle switch of gene regulatory network

Up to now, we have been studying the protein–protein interaction networks. The biological function in cells often involves active participation of genes in addition to proteins. Therefore we need to study gene regulatory networks with proteins performing the biological function and regulated by genes.

It is the purpose of this section to study the global robustness problem directly from the properties of the potential landscape for a simple yet important gene regulatory network: a toggle switch [20, 27, 50–55, 57–61]. Figure 4.14 shows a toggle switch. Gene networks often involve many degrees of freedom. To resolve the issue of multidimensionality, instead of using direct Monte Carlo simulation [27] for solving the master equations, a Hartree mean-field approximation can be applied to reduce the dimensionality and address the global issues [10, 20, 51, 59, 60].

There are three aims of this section. Our first aim is to develop a time dependent Hartree approximation scheme [20] to solve the associated master equations to follow the evolution of multidimensional probability of the network. Our second aim is to construct the underlying potential landscape for a toggle switch [61] and explore both the steady state and time evolution of the landscape. Our third aim is to study the phase diagram of the system and the kinetic timescale from one stable basin of attraction to another in different conditions. We will address the global robustness condition for a toggle switch.

As our goal is to uncover the potential landscape, we will first study the chemical reaction network involved in gene regulations. In particular we need to take into account the intrinsic statistical fluctuations due to the finite number of molecules in the cells. The statistical nature of the chemical reactions can be captured by the corresponding master equations. We will establish master equations for the gene regulations which describe the evolution of the networks probabilistically. The master equation is almost impossible to solve due to its inherent huge dimensions. We will therefore use the mean-field approximation to reduce the dimensionality [10, 20, 51]. In this way, we can follow the time evolution and steady-state probability of the protein concentrations. The steady-state probability is closely associated with the underlying potential energy landscape, which is our ultimate target.

Methods

Gene expression is regulated in various and complex ways, which can be represented by many coupled biochemical reactions. In this report our goal is not to just explain some specific gene network system as accurately as possible but to illustrate mathematical tools for exploring the general mechanisms of transcriptional regulatory gene networks. We will therefore take abstractions of some essential biochemical reactions from complicated reactions of diverse systems [20].

Let us start with the explanation of some terminology used in this manuscript: an *activator* is a regulatory protein that increases the level of transcription, a *repressor* is a regulatory protein that decreases the level of transcription. By *operator* we mean the DNA site or the gene where regulatory proteins (either an activator or a repressor) bind. First we are interested in the effect of *operator fluctuations* by which we mean the biochemical reactions that change the state of the operator. The operator is said to be in an occupied state if a regulatory protein is bound to it and in an unoccupied state otherwise. For the repressor we include the following reaction:

$$O_\alpha^1 + q_{\alpha\beta} \mathcal{M}_\beta \xrightarrow{h_{\alpha\beta}^R} O_\alpha^0 \tag{4.23}$$

$$O_\alpha^0 \xrightarrow{f_{\alpha\beta}^R} O_\alpha^1 + q_{\alpha\beta} \mathcal{M}_\beta \tag{4.24}$$

where $O_\alpha^{1(0)}$ stands for the active(inactive) operator state of gene α, $\mathcal{M}_\beta$ represents the regulatory protein synthesized or produced by gene β, and $q_{\alpha\beta}$ is for the multimer-type of proteins. For example, if $q_{AB} = 2(3)$, dimer(tetramer) proteins produced from gene B repress the expression of gene A; $h_{\alpha\beta}^R$ and $f_{\alpha\beta}^R$ are reaction probabilities per unit time. In the similar way, we may also consider the activator:

$$O_\alpha^0 + q_{\alpha\beta} \mathcal{M}_\beta \xrightarrow{h_{\alpha\beta}^P} O_\alpha^1 \tag{4.25}$$

$$O_\alpha^1 \xrightarrow{f_{\alpha\beta}^P} O_\alpha^0 + q_{\alpha\beta} \mathcal{M}_\beta. \tag{4.26}$$

Notice that the superscript 1(0) in $\mathcal{O}_\alpha^{1(0)}$ indicates the activity state of the operator and does not represent the bound state of regulatory protein. We will say the "gene is on(off)" when the operator of the gene is active(inactive). The gene will be "on" when it is occupied by activators or when repressors are unbound from it.

Next we include the transcription and translation steps. Here we ignore mRNA and consider only one step combining transcription and translation:

$$\emptyset \quad \xrightarrow{g_{\alpha 1}} \quad b_\alpha M_\alpha \quad \text{for} \quad \mathcal{O}_\alpha^1 \tag{4.27}$$

$$\emptyset \quad \xrightarrow{g_{\alpha 0}} \quad b_\alpha M_\alpha \quad \text{for} \quad \mathcal{O}_\alpha^0 \tag{4.28}$$

$$M_\alpha \quad \xrightarrow{k_\alpha} \quad \emptyset \tag{4.29}$$

where $\emptyset$ is used to denote a protein sink or source, b_α stands for the burst size of produced proteins (M_α), $g_{1(0)}$ is a protein synthesis probability per unit time, and k_α is a degradation probability per unit time.

We can say that (4.23–4.29) are "effective reactions" of the transcriptional regulatory gene network system. Roughly speaking we can say the other biochemical reactions could be taken into account by adjusting the parameters of the reaction probabilities per unit time. In this sense the reaction parameters are not really constants but functions of time. Furthermore, the proteins may not be well mixed in the cell, and the number of proteins could be a function of position. So we can generalize this formalism in a space-dependent manner. We also can add more species and reactions to the master equations. In this report we will assume homogeneity of the number of proteins and ignore the time delay (for example, due to the translation process) so that all the parameters are constants. Now we can construct the master equation based on the above assumptions and chosen "effective reactions."

Mean-field approximation of the master equation

The master equation is the equation for the time evolution of the probability of some specific state P [20]:

$$P(n_A, S_A, n_B, S_B, n_C, S_C, \cdots, t) \tag{4.30}$$

where A, B, C,.... is the label of each gene; $n_A, n_B, n_C, \cdots$ is the number of proteins expressed by gene A, B, C,... respectively. $S_A, S_B, S_C, \cdots$ is 1 or 0, and represents the activity state of the operator. The number of states, $\mathcal{N}$, is $n_A \times 2 \times n_B \times 2 \times n_C \times 2 \times \cdots$. We expect to have $\mathcal{N}$-coupled differential equations, which is not feasible to solve. Following a mean-field approach [10], we use Hartree mean-field approximation to split the probability into the products of individual ones: First, let us assume

$$P(n_A, S_A, n_B, S_B, n_C, S_C, \cdots, t) = \prod_\beta P(n_\beta, S_\beta, t), \tag{4.31}$$

and sum over all indexes except one specific index that we are interested in, say α. This effectively reduces the dimensionality from exponential $n_A \times n_B \times \cdots n_N \times 2^N$ to multiples $(n_A + n_B + \cdots) \times 2 \times N$ and therefore the problem is computationally tractable. Finally we are left with two equations for $P(n_\alpha, 1)$ and $P(n_\alpha, 0)$ (in fact, these are not just two equations because n_α varies from zero to hundreds). With the two component vector notation,

$$\mathbf{P}_\alpha := \left(\begin{array}{c} P_{\alpha 1}(n_\alpha, t) \\ P_{\alpha 0}(n_\alpha, t) \end{array} \right) := \left(\begin{array}{c} P(n_\alpha, S_\alpha = 1, t) \\ P(n_\alpha, S_\alpha = 0, t) \end{array} \right), \tag{4.32}$$

we have the compact form for the network:

$$\frac{\partial}{\partial t} \mathbf{P}_\alpha(n_\alpha, t) = \left(\begin{array}{cc} g_{\alpha 1} & 0 \\ 0 & g_{\alpha 0} \end{array} \right) \{ \theta(n_\alpha - b_\alpha)\mathbf{P}_\alpha(n_\alpha - b_\alpha, t) - \mathbf{P}_\alpha(n_\alpha, t) \}$$

$$+ k_\alpha \{ (n_\alpha + 1)\mathbf{P}_\alpha(n_\alpha + 1, t) - n_\alpha \mathbf{P}_\alpha(n_\alpha, t) \}$$

$$+ \left[\underbrace{\sum_\beta \left(\begin{array}{cc} -\mathcal{H}^R_{\alpha\beta} & f^R_{\alpha\beta} \\ \mathcal{H}^R_{\alpha\beta} & -f^R_{\alpha\beta} \end{array} \right)}_{\text{Repressive interaction}} + \underbrace{\sum_\beta \left(\begin{array}{cc} -f^P_{\alpha\beta} & \mathcal{H}^P_{\alpha\beta} \\ f^P_{\alpha\beta} & -\mathcal{H}^P_{\alpha\beta} \end{array} \right)}_{\text{Promoting interaction}} \right] \mathbf{P}_\alpha(n_\alpha, t).$$

$$\tag{4.33}$$

where

$$\mathcal{H}^{R(P)}_{\alpha\beta} := h^{R(P)}_{\alpha\beta} \underbrace{< \frac{n_\beta!}{(n_\beta - q_{\alpha\beta})! q_{\alpha\beta}!} >}_{=_{n_\beta} C_{q_{\alpha\beta}}}, \quad q_{\alpha\beta} := \left(\begin{array}{cc} 1 & \text{monomer} \\ 2 & \text{dimer} \\ 3 & \text{trimer} \\ q & \text{q-th multimer} \end{array} \right). \tag{4.34}$$

Notice that (4.33) is simply a "birth–death" process without the last term. We will call the first two terms in (4.33) "birth–death part" or "drift and diffusion part" in the view point of diffusional Fokker–Plank equation [50, 58]. Furthermore we will call the last term the "operator fluctuation part." In (4.33), all other indexes except α appear only in $\mathcal{H}_{\alpha\beta}$ in the ensemble-averaged form ($f_{\alpha\beta}$ is just some number). If we deal with the one-gene case, there is no ensemble average in (4.34). The first effect of the operator fluctuation is the sum over n_β and S_β. The second effect is to cancel out many of the "birth–death" terms of other genes. Since $\alpha = A,B,C,\cdots$, we have the vector equation set of the same numbers as those of the genes. They are coupled to each other through the term $\mathcal{H}_{\alpha\beta}$. All network interactions can be determined by assigning every $h_{\alpha\beta}$. b_α is the number of proteins produced in bursts from gene α, and θ is a step function. In (4.34) we take into account several kinds of binding proteins, and use proper combinatorics and ensemble average.

Quantum field theoretic description

The techniques of quantum field theory can be used to solve the master equation [10, 20, 59]. The first step is to construct a many-body "quantum state." Notice that the probabilities defined by (4.32) are imbedded in the quantum state as coefficients (4.36)

$$|\Psi\rangle = \prod_\alpha |\psi_\alpha\rangle = |\psi_A\rangle \otimes |\psi_B\rangle \otimes |\psi_C\rangle \otimes \cdots \tag{4.35}$$

where

$$|\psi_\alpha\rangle := \begin{pmatrix} \sum_{n_\alpha} P_{\alpha 1}(n_\alpha, t)|n_\alpha\rangle \\ \sum_{n_\alpha} P_{\alpha 0}(n_\alpha, t)|n_\alpha\rangle \end{pmatrix}. \tag{4.36}$$

In (4.35) we make an ansatz of Hartree-type product for the many-body state. Then non-Hermitian "Hamiltonian" of only repressive proteins, Ω, yields:

$$\Omega := \sum_\alpha \left(\left(\mathbf{D}_\alpha + \begin{pmatrix} -\mathbf{H}_{\alpha\alpha} & 0 \\ \mathbf{H}_{\alpha\alpha} & 0 \end{pmatrix} \right) \otimes \underbrace{\prod_{\beta \neq \alpha} 1_\beta + 1_\alpha \otimes \sum_{\beta \neq \alpha} \begin{pmatrix} -\mathbf{H}_{\alpha\beta} & 0 \\ \mathbf{H}_{\alpha\beta} & 0 \end{pmatrix} \otimes \prod_{\gamma \neq \alpha, \beta} 1_\gamma}_{\text{interaction part (network)}} \right)$$

$$\underbrace{\phantom{\left(\mathbf{D}_\alpha + \begin{pmatrix} -\mathbf{H}_{\alpha\alpha} & 0 \\ \mathbf{H}_{\alpha\alpha} & 0 \end{pmatrix} \right)}}_{\text{self-repressive}}$$

$$\tag{4.37}$$

where

$$\mathbf{D}_\alpha := \begin{pmatrix} g_{\alpha 1}((a_\alpha^\dagger)^{b_\alpha} - 1) + k_\alpha(a_\alpha - a_\alpha^\dagger a_\alpha) & f_\alpha \\ 0 & g_{\alpha 0}((a_\alpha^\dagger)^{N_\alpha} - 1) + k_\alpha(a_\alpha - a_\alpha^\dagger a_\alpha) - f_\alpha \end{pmatrix} \tag{4.38}$$

$$\mathbf{H}_{\alpha\beta} := \sum_{q_{\alpha\beta}} h_{\alpha\beta} \frac{(a_\beta^\dagger a_\beta)!}{(a_\beta^\dagger a_\beta - q_{\alpha\beta})! q_{\alpha\beta}!}. \tag{4.39}$$

For each protein concentration, a creation and an annihilation operator are introduced, such that $a^\dagger|n\rangle = |n+1\rangle$ and $a|n\rangle = n|n-1\rangle$. These operators satisfy $[a, a^\dagger] = 1$. The generalization to include activating proteins is straightforward. While the "state vector" is a simple product of individual genes, the operator product form of Ω is chosen deliberately to reproduce the original master equation (4.33). The Ω of a many gene system seems to be $\Omega = \sum \Omega_i$ [10], but it would not be the simple sum of individual operators because of the interaction terms. Like the master equation, $\mathbf{D}_\alpha$ is the "birth–death part" and plays a role of diffusion and drift terms in the context of Fokker–Plank equation. The second term and third term in Eq.(4.37) are repressor-related terms and $\mathbf{H}_{\alpha\beta}$ is the counter part of $\mathcal{M}_{\alpha\beta}$ (4.34). Finally we have the following quantum system:

$$\frac{\partial}{\partial t}|\Psi\rangle = \Omega|\Psi\rangle. \tag{4.40}$$

To complete the mean-field approximation, we need to average all interaction effects by doing an "inner product" with some "reference" state, which is a two component generalization of the Glauber state [59]. If we are interested in an α-gene (operator) state and the associated protein, we may define a "reference" state:

$$\langle \tilde{\phi}_\alpha | := \prod_{\beta \neq \alpha} \langle S_\beta | := \prod_{\beta \neq \alpha} \Big(\langle 0 | e^{a_\beta} \quad \langle 0 | e^{a_\beta} \Big). \tag{4.41}$$

Then

$$\left\langle \tilde{\phi}_\alpha \left| \frac{\partial}{\partial t} \right| \Psi \right\rangle = \langle \tilde{\phi}_\alpha | \Omega | \Psi \rangle \tag{4.42}$$

is equivalent to the master equation (4.33).

Ritz's variational method with coherent state ansatz

We will use the Rayleigh–Ritz variational method to obtain an approximate solution of a non-Hermitian Hamiltonian system (nonequilibrium system) like (4.40) [10, 20, 60]. The master equation is equivalent to the functional variation $\delta\Gamma/\delta\Phi = 0$ of an "effective action" $\Gamma = \int dt < \Phi|(\partial_t - \Omega)|\Psi >$. The method is analogous to the traditional procedure in quantum mechanics except the modification due to non-Hermitian properties of the operator, of which left eigenvectors and right eigenvectors do not have to be the same. We will make a ket state ansatz, $|\Psi\rangle$, and a bra state ansatz, $\langle\Phi|$, respectively [10]:

$$|\Psi\rangle := \prod_\alpha |\psi_\alpha\rangle \quad \langle\Phi| := \prod_\alpha \langle\phi_\alpha| \tag{4.43}$$

$$|\psi_\alpha\rangle := \begin{pmatrix} C_{\alpha 1} e^{X_{\alpha 1}(a_\alpha^\dagger - 1)}|0_\alpha\rangle \\ C_{\alpha 0} e^{X_{\alpha 0}(a_\alpha^\dagger - 1)}|0_\alpha\rangle \end{pmatrix} \tag{4.44}$$

$$\langle\phi_\alpha| := \Big(\langle 0_\alpha | e^{a_\alpha} e^{\sigma_{\alpha 1} + \lambda_{\alpha 1} a_\alpha} \quad \langle 0_\alpha | e^{a_\alpha} e^{\sigma_{\alpha 0} + \lambda_{\alpha 0} a_\alpha} \Big) \tag{4.45}$$

where $C_{\alpha 1}, C_{\alpha 0}, X_{\alpha 1}, X_{\alpha 0}, \alpha_{\alpha 1}, \alpha_{\alpha 0}, \lambda_{\alpha 1}, \lambda_{\alpha 0}$ are time-dependent parameters to be determined by the variational principle. The ket ansatz is chosen as coherent state, which corresponds to Poisson distribution. $C_{\alpha 1}$ and $C_{\alpha 0}$ are the probabilities of the two DNA-binding states $S = 1$ and $S = 0$, respectively. The coupled dynamics of the DNA-binding state and the protein distribution is described as the motion of wavepackets with amplitudes $C_{\alpha 1}$ and $C_{\alpha 0}$ as well as by means of the protein concentrations at $X_{\alpha 1}$ and $X_{\alpha 0}$ (from Poisson distribution ansatz). With the following notation

$$\alpha_{\alpha j}^R = C_{\alpha 1}, C_{\alpha 0}, X_{\alpha 1}, X_{\alpha 0} \quad j = 1, 2, 3, 4 \quad \text{respectively} \quad \alpha = A, B, C, . \tag{4.46}$$

$$\alpha_{\alpha j}^L = \alpha_{\alpha 1}, \alpha_{\alpha 0}, \lambda_{\alpha 1}, \lambda_{\alpha 0} \quad j = 1, 2, 3, 4 \quad \text{respectively} \quad \alpha = A, B, C, . \tag{4.47}$$

Here, $< \Phi|(\alpha^L = 0)$ is set to be consistent with the probabilistic interpretation $< \Phi(\alpha^L = 0)|\Psi(\alpha^R) >= 1$. The condition for the extremum of the action with respect to $< \Phi|$ is:

$$\left\langle \frac{\partial \Psi}{\partial \alpha^L_{\alpha i}} \Big| \frac{\partial \Psi}{\partial \alpha^R_{\alpha j}} \right\rangle \dot{\alpha}^R_{\alpha j} = \left\langle \frac{\partial \Psi}{\partial \alpha^L_{\alpha i}} |\Omega| \Psi \right\rangle, \qquad \text{all } \alpha^L_{\alpha i} = 0 \tag{4.48}$$

which is reduced to the coupled ordinary differential equations with parameters:

$$\dot{C}_{\alpha 1} = -C_{\alpha 1} H^R_{\text{total}} + C_{\alpha 0} f^R_{\text{total}} + C_{\alpha 0} H^P_{\text{total}} - C_{\alpha 1} f^P_{\text{total}} \tag{4.49}$$

$$\dot{C}_{\alpha 0} = -\dot{C}_{\alpha 1} \tag{4.50}$$

$$C_{\alpha 1} \dot{X}_{\alpha 1} + X_{\alpha 1} \dot{C}_{\alpha 1}$$
$$= C_{\alpha 1}(b_\alpha g_{\alpha 1} - k_\alpha X_{\alpha 1})$$
$$\quad - C_{\alpha 1} H'^R_{\text{total}} + C_{\alpha 0} f^R_{\text{total}} X_{\alpha 0} + C_{\alpha 0} H'^P_{\text{total}} - f^P_{\text{total}} C_{\alpha 1} X_{\alpha 1} \tag{4.51}$$

$$C_{\alpha 0} \dot{X}_{\alpha 0} + X_{\alpha 0} \dot{C}_{\alpha 0}$$
$$= C_{\alpha 0}(b_\alpha g_{\alpha 0} - k_\alpha X_{\alpha 0})$$
$$\quad + C_{\alpha 1} H'^R_{\text{total}} - C_{\alpha 0} f^R_{\text{total}} X_{\alpha 0} - C_{\alpha 0} H'^P_{\text{total}} + f^P_{\text{total}} C_{\alpha 1} X_{\alpha 1} \tag{4.52}$$

$$H^R_{\text{total}} := \langle 0|e^{a_\alpha} \mathbf{H}^R_{\alpha\alpha} e^{X_{\alpha 1}(a^\dagger_\alpha - 1)}|0\rangle + \sum_{\beta \neq \alpha} \langle \tilde{\phi}_\beta|\mathbf{H}^R_{\alpha\beta}|\psi_\beta\rangle \tag{4.53}$$

$$H^P_{\text{total}} := \langle 0|e^{a_\alpha} \mathbf{H}^P_{\alpha\alpha} e^{X_{\alpha 0}(a^\dagger_\alpha - 1)}|0\rangle + \sum_{\beta \neq \alpha} \langle \tilde{\phi}_\beta|\mathbf{H}^P_{\alpha\beta}|\psi_\beta\rangle \tag{4.54}$$

$$f^{R(P)}_{\text{total}} := f^{R(P)}_{\alpha\alpha} + \sum_{\beta \neq \alpha} f^{R(P)}_{\alpha\beta} \tag{4.55}$$

$$H'^R_{\text{total}} := \langle 0|e^{a_\alpha} a^\dagger_\alpha \mathbf{H}^R_{\alpha\alpha} e^{X_{\alpha 1}(a^\dagger_\alpha - 1)}|0\rangle + \sum_{\beta \neq \alpha} \langle \tilde{\phi}_\beta|\mathbf{H}^R_{\alpha\beta}|\psi_\beta\rangle)X_{\alpha 1} \tag{4.56}$$

$$H'^P_{\text{total}} := \langle 0|e^{a_\alpha} a^\dagger_\alpha \mathbf{H}^P_{\alpha\alpha} e^{X_{\alpha 0}(a^\dagger_\alpha - 1)}|0\rangle + \sum_{\beta \neq \alpha} \langle \tilde{\phi}_\beta|\mathbf{H}^P_{\alpha\beta}|\psi_\beta\rangle)X_{\alpha 1} \tag{4.57}$$

where $\langle \tilde{\phi}_\beta|$ is defined by (4.41). We can interpret (4.49) as a master equation of probabilities of active operator states related to reactions (4.23–4.26). However its change of the rate is based on the "average" number of proteins (4.53–4.54) unlike usual master equations which deal with the specific state. (4.50) is just probability conservation. The first term of (4.51) and (4.52) is related to protein synthesis and degradation (4.27–4.29) and the rest are the effects of operator fluctuations.

Equations (4.49–4.52) are the general equations for parameters; we can use these equations for any number of gene systems and for any kind of regulatory patterns by assigning the binding and unbinding probability rate.

Interpretation of the solutions

The final output we get from the equations are basically moments. From these moments we need to construct the total probability [20]. There are several important features to be pointed out. We start with the single gene case.

First, notice that the total probability does not have the structure of $C_1 P_1 + C_0 P_0$. We start with a two-component column vector and in order to extract the physical observable we need to do the inner product with a two-component row state vector. (We never add spin up and down components directly in quantum mechanics.) The total probability should therefore not follow the steps of constructing P_1 and P_0 first and then weighting by C_1 and C_0. The correct procedure is the following. With the moments, the solutions of equations, we construct new moments:

$$1 = C_1 + C_0$$

$$<n> = C_1 <n>_{on} + C_0 <n>_{off}$$

$$<n^k> = C_1 <n^k>_{on} + C_0 <n^k>_{off} \qquad k = 2, 3, \cdots$$

In principle we can get arbitrary order of moments and construct the corresponding probability if the equations are closed. In practice, however, we may choose one of two probability distributions: Poisson or Gaussian distributions.

Second, the probability obtained above corresponds to one limit point or basin of attraction. One solution of the equations determines one of the limit points and gives also the variation around the basin of attraction, so it is intrinsic. If the system allows multistability then there are several probability distributions localized at each basin of attraction, but with different variations. Thus the total probability is the weighted sum of all these probability distributions. The weighting factors (w^a, w^b) are the size of basin, which is nothing but the relative size of the set of initial values ending up with a specific basin of attraction.

$$P(n) = w^a P^a(n) + w^b P^b(n) \qquad w^a + w^b = 1. \tag{4.58}$$

Notice that the steady-state solution is not enough to describe the total probability. It does not say anything about the *volume* of basin, only says the limit *point*. So the efforts to derive an effective "potential energy" from the steady-state solution on general grounds need to take into account the volume of the basin of attraction. One simple exception is the *symmetric* toggle switch, where the weighting factors are simply $(0.5, 0.5)$ by symmetry.

Third, the total probability of many genes is simply the product of each gene based on our basic assumption, the mean-field approximation. For example, the probability of a toggle switch can be written as

$$P(n_A, n_B) = w^a P^a(n_A) P^a(n_B) + w^b P^b(n_A) P^b(n_B) \tag{4.59}$$

where a and b denote each limit point and w^a, w^b are the weighting factors. Even though it is simply multiplication, the interactions between them are already taken into account from the coupled equations.

Finally, once we have the total probability we can construct the potential energy (or potential landscape) by the relationship with the steady-state probability:

$$U(\vec{n}) = -\ln P(\vec{n}). \tag{4.60}$$

This is the reverse order of the usual statistical mechanics of first obtaining the potential energy function, exponentially Boltzmann weighting it, and then studying the partition function or probability of the associated system. Here we look for the inherent potential energy function from the steady-state probability. In the gene-network system, every chemical parameter such as protein production/decay rates, and binding/unbinding rates will contribute to the fluctuation of the system. All these effects are encoded in the total probability distribution, consequently in the underlying potential energy landscape.

Results

We will look at an important example of two genes mutually interacting with each other. The interactions are through the proteins synthesized by the genes which will act back to regulate the gene switch. The bacterial λ phage is a good biological example of a toggle switch. The two lysogenic and lysosic genes are both stable and robust. It has been a long-standing problem to explain why the lambda phage is so stable [10, 13, 20, 50, 51, 57, 62]. We will address this issue in the presence of the intrinsic statistical fluctuations of the finite number of the proteins in the cell by exploring the underlying potential energy landscape [20]. First we will solve the master equation to obtain both the time-dependent and steady-state probability distributions of the protein concentrations of the corresponding genes. Then we will infer the underlying potential energy landscape from the steady-state probability distribution of the protein concentrations. We will consider the symmetric toggle switch.

All applications to specific network systems start with (4.49–4.52). First, we choose the number of genes and design the interaction type (network topology) and protein types (multimers). Second, we assign the strength of parameters. Then we can solve this coupled ordinary differential equation (ODE) system numerically with certain initial conditions. We will consider a toggle switch [61] of two genes as shown in Fig. 4.14 which has wide application in molecular biology such as the bacteria lambda phage problem [57]. Let us start with the toggle switch case [20].

Fig. 4.14 Toggle switch: protein A (dotted line) represses gene B, protein B (solid line) represses gene A.

Symmetric toggle switch

For the symmetric switch, we first solve the equations of motion determining the amplitude, the mean, and higher-order moments of the probability distribution of the protein concentrations of the corresponding genes [20]. We can use the moments to infer the corresponding probability distribution. These are given below.

Poisson and moment equation solutions of master equations

We can solve the master equation with two methods. One is the Poisson ansatz mentioned above by assuming the inherent Poisson distribution and the other is the exact method using the moment equation. For the inherent Poisson distribution, we can write down the equations of motion for the amplitude and mean Poisson ansatz:

$$\dot{C}_{A1} = -\frac{h_A}{2} C_{A1} \{C_{B1} X_{B1}^2 + (1 - C_{B1}) X_{B0}^2\} + f_A (1 - C_{A1}) \tag{4.61}$$

$$\dot{X}_{A1} = -f_A \frac{1 - C_{A1}}{C_{A1}} (X_{A1} - X_{A0}) + g_{A1} - k_A X_{A1} \tag{4.62}$$

$$\dot{X}_{A0} = f_A (X_{A1} - X_{A0}) + g_{A0} - k_A X_{A0} - \frac{X_{A1} - X_{A0}}{1 - C_{A1}} \dot{C}_{A1} \tag{4.63}$$

$$\text{three more equations } (A \leftrightarrow B) \tag{4.64}$$

where we eliminated two variables by the probability conservation ($C_{\alpha 1} + C_{\alpha 0} = 1$), and recollected terms.

For the exact solution with moment equations, we can also write down the equations of motion of the moment of protein concentration of the corresponding genes. The corresponding moment equations:

$$\dot{C}_{A1} = -\frac{h_A}{2} C_{A1} \{C_{B1} (<n_{B1}^2> - <n_{B1}>) + (1 - C_{B1})(<n_{B0}^2> - <n_{B0}>)\} + f_A (1 - C_{A1}) \tag{4.65}$$

$$<\dot{n}_{A1}> = -f_A \frac{1 - C_{A1}}{C_{A1}} (<n_{A1}> - <n_{A0}>) + g_{A1} - k_A <n_{A1}> \tag{4.66}$$

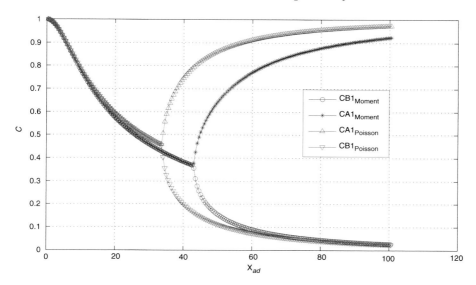

Fig. 4.15 Probability C that genes are in the active state as a function of $X_{ac} = (g_1 + g_0)/2k_A$ for a symmetric switch showing the bifurcation: exact moment equation solutions are compared with Poisson ansatz solutions. $0 < X_{ad}(= g_1/2) < 100, K_A = k_B = 1, f_A = f_B = 0.5, h_A = h_B = f_A/500,$ and $g_{A0} = g_{B0} = 0.$

$$<\dot{n}_{A0}> = f_A(<n_{A1}> - <n_{A0}>) + g_{A0} - k_A <n_{A0}> - \frac{<n_{A1}> - <n_{A0}>}{1 - C_{A1}} \dot{C}_{A1} \qquad (4.67)$$

$$\dot{C}_{A1} <n_{A1}^2> + C_{A1} <\dot{n}_{A1}^2> \qquad (4.68)$$
$$= C_{A1}g_{A1}(2 <n_{A1}> +1) + k_A C_{A1}(-2 <n_{A1}^2> + <n_{A1}>)$$
$$+ f_A(1 - C_{A1}) <n_{A0}^2> - C_{A1}\frac{h_A}{2} <n_{A1}^2> (C_{B1}(<n_{B1}^2> - <n_{B1}>)$$
$$+ (1 - C_{B1})(<n_{B0}^2> - <n_{B0}>))$$

$$-\dot{C}_{A1} <n_{A0}^2> + (1 - C_{A1})<\dot{n}_{A0}^2> \qquad (4.69)$$
$$= (1 - C_{A1})g_{A0}(2 <n_{A0}> +1) + k_A(1 - C_{A1})(-2 <n_{A0}^2> + <n_{A0}>)$$
$$- f_A(1 - C_{A1}) <n_{A0}^2> + C_{A1}\frac{h_A}{2} <n_{A1}^2> (C_{B1}(<n_{B1}^2> - <n_{B1}>)$$
$$+ (1 - C_{B1})(<n_{B0}^2> - <n_{B0}>))$$

five more equations $(A \leftrightarrow B)$. $\qquad (4.70)$

Mono- versus bi-stability for symmetric toggle switch

By giving some initial conditions, and taking the long time limit, we can obtain the steady-state solution [20]. Moment equation solutions can be used to infer the corresponding steady probabilities. We fix all parameters except the protein synthesis rate $g_{A1}(= g_{B1})$. We look at the probability of genes that are in the active state versus essentially the relative importance of synthesis rate versus degradation rate. By increasing the synthesis rate g_{A1} we can observe the bifurcation from mono-stable state to the bi-stable state passing a certain critical point. Figure 4.15 is the illustration of the result by taking the long time limit of the equations of motion. Two curves (with subscript "Moment")

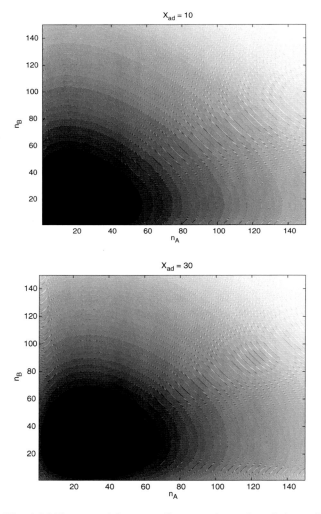

Fig. 4.16 The potential energy of symmetric toggle switch as a function of the numbers of protein A and the numbers of protein B for different $X_{ad} = g_{A1}/2k_A$. The other parameters are the same as Fig. 4.15. (See plate section for color version.)

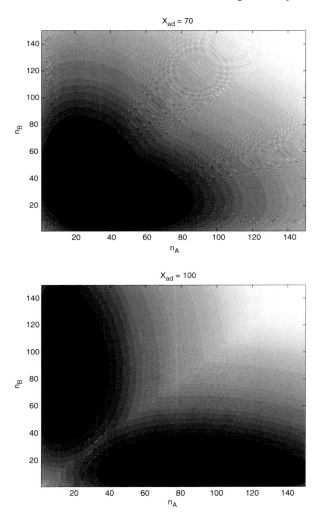

Fig. 4.16 (Cont.)

are from moment equations and the others are from the Poisson ansatz. This is consistent with the results directly from time-independent equations (Fig. 4.3(a) in [51]). We use the parameter $X_{ad} = \bar{g}/k$ as the horizontal axis variable. It is simply $g_1/2$ in our choice.

In the parameter range in which the bi-stability occurs, we found two limit points (named a and b) in the numerical analysis. (All illustrations in this chapter are based on the Poisson ansatz for simplicity, but it can be easily done with the moment equations. Qualitative features will not be changed.) Now from the solution of the equations, we can construct the probability of protein numbers or concentrations [20]:

$$P(n_A, n_B) = w^a P^a(n_A) P^a(n_B) + w^b P^b(n_A) P^b(n_B). \tag{4.71}$$

For the symmetric toggle switch case, the weight factor is simply $(0.5, 0.5)$ due to symmetry.

Potential energy landscape: mono-stability to bi-stability

The steady-state probability can be inferred from the related moments. As we discussed, the steady-state distribution function $P(\vec{n})$ for the state variable $\vec{n}$ can be expressed to be exponential in a function $U(\vec{n})$:

$$P(\vec{n}) = \exp\{-U(\vec{n})\}, \tag{4.72}$$

where $P(\vec{n})$ is already normalized. From the steady-state distribution function, we can therefore identify U as the generalized potential energy function of the network system [20]. In this way, we map out the potential landscape. Figure 4.16 shows the potential landscape.

We can see that when the protein synthesis rate is small relative to degradation rate, only one single basin of attraction exists for the underlying potential landscape. For large enough protein synthesis rate relative to degradation rate, two basins of attraction emerge. Once we have the potential landscape, we can discuss the global stability of the gene regulatory networks. The timescale of the transition between the two stable minimum basins of attraction can be estimated by $\tau \sim \tau_0 \exp[U^{\neq} - U_{min}]$ [20, 63]. Here τ_0 is the pre-factor and τ is the timescale of transition from one basin of attraction to the other. $U^{\neq}$ is the potential energy at the saddle point between the two stable basins of attraction. U_{min} is the potential energy at one of the basins of attraction. Thus $U^{\neq} - U_{min}$ represents the potential energy barrier height between two stable basins of attraction. In Fig. 4.17 we can see as the synthesis rate and unbinding rate of protein to DNA increase relative to the degradation rate, the potential barrier height between the two basins of attraction increases. The timescale of transition from one basin of attraction to the other exponentially increases with the barrier height.

Figure 4.17 shows the phase diagram of the parameter ranges for the mono-stable basin and two bi-stable basins of attraction. We can see when the synthesis rate and unbinding rate of protein to DNA are low relative to the degradation rate, the potential landscape prefers one stable basin of attraction. As the synthesis rate and unbinding rate of protein to DNA increases relative to the degradation rate, the potential landscape gradually develops the two stable

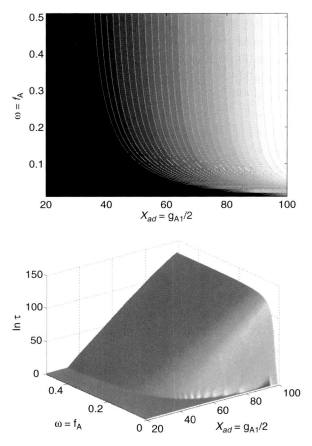

Fig. 4.17 The timescale of the transition between the two stable minimum basins of attraction as a function of $\omega = f_A/k_A$ and $X_{ad} = g_{A1}/2k_A$. The other parameters are the same as Fig. 4.15.

basins of attraction from the mono-stable one. There is a transition from mono-stable to bi-stable basins of attraction of the underlying potential landscape at certain parameters.

This is an illustration of how biological robustness is realized for the toggle switch. As the protein synthesis rate and unbinding rate of protein to DNA increase relative to the degradation rate, more proteins are synthesized. These proteins are strong repressors. This leads to smaller fluctuations. Furthermore, the associated barrier height between the two basins of attraction also becomes large, then the two basins of attraction become more stable since it is harder to go from one basin to the other. So small fluctuations and large barrier heights both serve as the source for the robustness and stability of the gene toggle switch. In other words it is more unlikely for the system to change from its one basin of attraction to the other. Therefore the system becomes robust. The robustness issue is not yet well understood for cellular networks in general. Here we explore the robustness of the switches against the intrinsic statistical fluctuations coming from the finite number of proteins and DNAs. This is clearly very important and has potential applications to the robustness problem of phage λ in bacteria.

Discussions

Finding the multidimensional potential landscape is the key to address the important global issues such as robustness of the cellular networks. We have uncovered the underlying potential landscape of a simple gene network: toggle switch. We have found as the protein synthesis rate and the unbinding of protein to DNA rate relative to degradation change from small to large, the underlying potential landscape changes from having mono-stable to bi-stable basins of attraction. These basins correspond to the stable biologically functional states. The potential barrier between the two basins determines the timescale of conversion from one to the other. We found as the protein synthesis rate and unbinding of protein with DNA rate relative to degradation become greater, the potential energy barrier becomes greater and furthermore the statistical fluctuations are effectively more severely suppressed. This leads to the robustness of the biological basins of the gene switches.

In principle, our approach can be generalized to more realistic networks involving multiple genes as well as additional levels of regulation. This could be realized by averaging the interactions among genes in the corresponding master equations. It effectively reduces the dimensionality of the problem from exponential to polynomial number of degrees of freedom. It is worthwhile to note the limitation of this approach. When the interactions among genes are very strong, our approach is less effective.

Recently, the synthetic biology becomes an important part of the systems biology [64–67]. There are significant progresses in this field. However, there still seems lacking of general principles and algorithms guiding the design and construction of synthetic gene networks. The robustness condition (see Fig. 4.17) found in this study would help us to identify the parameter and connectivity region to reach global robustness and function of the network. The optimal network design will be based on that. Furthermore, we can vary the parameters and connections to design different distinct features while maintaining the stability for the network.

The adaptive landscape idea was first introduced into biology by S. Wright in 1932 [68–71]. Landscape construction for one dimension is rather straightforward. However, even the two-dimensional case becomes nontrivial. The recent efforts in understanding globally the systems biology need the concept of land-scape. Progress has been made towards this from the dynamic system point of view where the nontrivial nature of a low-dimensional system was illustrated [17, 18, 20, 21, 72]. There are still conceptual and methodology issues remaining for high-dimensional systems. The stochastic method introduced here may pave the road towards solving this problem.

The model presented here can be modified to include more biochemical reactions. If we are interested in the role of mRNA, then we can consider transcription and translation processes separately. Or if we want to focus on the statistical fluctuations of on and off of the genes, then it is possible to generalize the formalism to compute the statistical fluctuations quantitatively. We also can take into account the spacial variation of the state variables such as the number of proteins.

Acknowledgments

JW would like to thank Professor P. G. Wolynes, Professor J. N. Onuchic, Professor P. Ao, Dr. X. M. Zhu, Professor. H. Qian, and Professor. Z. G. Wang for helpful discussions. JW would like to thank financial supports from NSF, ACS-PRF and NSF of China. JW would also like to thank the International Workshop on Bionetworks in July 2004 at Lijiang, China, where the idea of a funneled landscape for cellular networks was first presented by JW.

References

[1]   Davidson, E. H. *et al.* 2002. A genomic regulatory network for development. *Science* **295**, 1669–1673.

[2]   Huang, C. Y. & J. E. Ferrell Jr. 1996. Ultrasensitivity in the mitogen-activated protein kinase cascade. *Proc. Natl Acad. Sci.* **93**, 10 078–10 082.

[3]   Kholodenko, B. N. 2000. Negative feedback and ultrasensitivity can bring about oscillations in the mitogen-activated protein kinase cascades. *Eur. J. Biochem.* **267**, 1583–1593.

[4]   Ideker, T. *et al.* 2001. Integrated genomic and proteomic analyses of a systematically perturbed metabolic network *Science* **292**, 929–933.

[5]   Jeong, H., B. Tombor, R. Albert, Z. N. Oltvai, & A. L. Barabasi. 2000. The large-scale organization of metabolic networks. *Nature* **407**, 651–654.

[6]   Barabasi, A. L. & E. Bonabeau. 2003. Scale-free networks. *Sci. Am.* **288**, 60–69.

[7]   Maslov, S. & K. Sneppen. 2002. Specificity and stability in topology of protein networks. *Science* **296**, 910–913.

[8]   Milo, R., S. Shen-Orr, S. Itzkovitz, *et al.* 2002. Network motifs: simple building blocks of complex networks. *Science* **298**, 824–827.

[9]   Carlson, J. M. & J. Doyle. 2002. Complexity and robustness. *Proc. Natl Acad. Sci. USA* **99**, 2538–2544.

[10]   Sasai, M. & P. G. Wolynes. 2003. Stochastic gene expression as a many body problem. *Proc. Natl Acad. Sci. USA* **100**, 2374–2379.

[11]   Li, F., T. Long, Y. Lu, Q. Ouyang, & C. Tang. 2004. The yeast cell cycle network is robustly designed. *Proc. Natl Acad. Sci. USA* **101**, 4781–4786.

[12]   Ao, P. 2004. Potential in stochastic differential equations: novel construction. *J. Phys. A Math. Gen.* **37**, L25–L30.

[13]   Zhu, X. M., L. Lan, L. Hood, & P. Ao. 2004. Calculating biological behaviors of epigenetic states in the phage lambda life cycle. *Funct. Integr. Genomics* **4**, 188–195.

[14]   Qian, H. & D. A. Bear. 2005. Thermodynamics of stoichiometric biochemical networks far from equilibrium. *Biophys. Chem.* **114**, 213–220.

[15]   Qian, H. & T. C. Reluga. 2005. Nonequilibrium thermodynamics and nonlinear kinetics in a cellular signaling switch. *Phys. Rev. Lett.* **94**, 028101-4.

[16]   Hornos, J. E. M., D. Schultz, G. C. P. Innocentini, *et al.* 2005. Self-regulating gene: an exact solution. *Phys. Rev. E.* **72**, 051 907-1–051 907-5.

[17]   Wang, J., B. Huang, X. F. Xia, & Z. R. Sun. 2006. Funeled landscape leads to robustness of cellular network: MAP kinase signal transduction. *Biophys. J. Lett.* **91**, L54–L57.

[18]   Wang, J., B. Huang, X. F. Xia, & Z. R. Sun. 2006. Funneled landscape leads to robustness of cell networks: yeast cell cycle. *PLoS Comp. Biol.* 2, e147, 1385.

[19]   Kim, K., D. Lepzelte, & J. Wang. 2007. Single molecule dynamics, statistical fluctuations, amplitude and period of oscillations of a gene regulatory network: repressilator. *J. Chem. Phys.* **126**, 034 702.

[20]   Kim, K. & J. Wang. 2007. Potential energy landscape and robustness of a gene regulatory network: toggle switch. *PLoS Comp. Biol.* **3**, e60.

[21]   Wang, J. & B. Han. 2007. Quantifying robustness and dissipation cost of cell cycle network: funneled energy landscape perspectives. *Biophys. J.* **92**, 3755.

[22]   Kauffman, S. A. 1971. Gene regulation networks: a theory for their global structure and behavior. *Curr. Topics Dev. Biol.* **6**, 145–165.

[23]   Lotkar, A. J. 1925. *Elements of Physical Biology*. Baltimore, MD: Williams & Wilkins.

[24]   Volterra, V. 1926. Variations and fluctuations of the number of individuals in animal species living together. *Mem. Acad. Lincei* **2**, 31–113.

[25]   Abbott, L. F. 1994. Decoding neuronal firing and modeling neural networks. *Q. Rev. Biophys.* **27**, 291–331.

[26]   Levines, R. 1969. Some demographic and genetic conseqeuences of environmental heterogeneity for biological control. *Bull. Entomol. Soc. Amer.* **15**, 237–240.

[27]   Gillespie, D. T. 1977. Exact stochastic simulation of coupled chemical reactions. *J. Phys. Chem.* **81**, 2340–2361.

[28]   McAdams, H. H. & A. Arkin. 1997. Stochastic mechanisms in gene expression. *Proc. Natl Acad. Sci. USA* **94**, 814–819.

[29]   Elowitz M. B. & S. Leibler. 2000. A synthetic oscillatory network of transcriptional regulators. *Nature* **403**, 335–338.

[30]   Swain, P. S., M. B. Elowitz, & E. D. Siggia. 2002. Intrinsic and extrinsic contributions to stochasticity in gene expression. *Proc. Natl Acad. Sci. USA* **99**, 12 795–12 800.

[31]   Thattai, M. & A. van Oudenaarden. 2001. Intrinsic noise in gene regulatory networks. *Proc. Natl Acad. Sci. USA* **98**, 8614–8619.

[32]   Vilar, J. M.G., C. C. Guet, & S. Leibler. 2003. Modeling network dynamics: the *lac* operon, a case study. *J. Cell Biol.* **161**, 471–476.

[33]   Paulsson, J. 2004. Summing up the noise in gene networks. *Nature* **427**, 415–418.

[34]   Tyson, J. J., K. Chen, & B. Novak. 2001. Network dynamics and cell physiology. *Nat. Rev. Mol. Cell Biol.* **2**, 908–916.

[35]   Novak, B. & J. J. Tyson. 1997. Modeling the control of DNA replication in fission yeast. *Proc. Natl Acad. Sci. USA* **94**, 9147–9152.

[36]   Tyson, J. J. 1991. Modeling the cell division cycle: cdc2 and cyclin interactions. *Proc. Natl Acad. Sci. USA* **88**, 7328–7332.

[37]   Chen, K. C., L. Calzone, A. Csikasz-Nagy, *et al.* 2004. Integrative analysis of cell cycle control in budding yeast. *Mol. Bio. Cell.* **15**, 3841–3862.

[38]   Austin, R. H., K. Beeson, L. Eisenstein, *et al.* 1975. Ligand binding to myoglobin. *Biochemistry* **14**, 5355–5373.

[39]   Frauenfelder, H., S. G. Sligar, & P. G. Wolynes. 1991. The energy landscapes and motions of proteins. *Science* **254**, 1598–1603.

[40]   Wolynes, P. G., J. N. Onuchic, & D. Thrumalai. 1995. Navigating the folding routes. *Science* **267**, 1619–1622.

[41]   Wang, J. & G. M. Verkhivker. 2003. Energy landscape theory, funnels, specificity and optimal criterion of biomolecular binding. *Phys. Rev. Lett.* **90**, 188 101–4.

[42]   Van Kampen, N. G. 1992. *Stochastic Processes in Physics and Chemistry.* New York: Elsevier.

[43]   Freedman, D. 1983. *Markov Chains.* Berlin: Springer-Verlag.

[44]   Davis, M. H. A. 1993. *Markov Models.* London: Chapman & Hall.

[45]   Zhang, Y., M. Qian, Q. Ouyang, *et al.* 2006. Stochastic model of yeast cell cycle network. *Physica D* **219**, 35–39.

[46]   Hopfield, J. J. 1982. Neural networks and physical systems with emergent collective computational abilities. *Proc. Natl Acad. Sci. USA* **79**, 2554–2558.

[47]   de Groot, S. R. & P. Mazur. 1984. *Non-Equilibrium Thermodynamics.* New York: Dover Publications.

[48]   Schnakenberg, J. 1976. Network theory of microscopic and macroscopic behavior of master equation systems. *Rev. Mod. Phys.* **48**, 571–588.

[49]   Torres, N. V. 1994. Modeling approach to control of carbohydrate metabolism during citric acid accumulation by *Aspergillus niger*: I. Model definition and stability of the steady state. *Biotech. Bioeng.* **44**, 104–110

[50]   Bialek, W. 2003. Stability and noise in biochemical switches *Adv. Neural Inform. Process.* **13**, 103–109.

[51]   Walczak, A. M., M. Sasai, & P. G. Wolynes. 2005. Self consistent proteomic field theory of stochastic gene switches. *Biophys. J.* **88**, 828–850.

[52]   Hasty, J., J. Pradines, M. Dolnik, & J. J. Collins. 2000. Noise-based switches and amplifiers for gene expression , *Proc. Natl Acad. Sci. USA* **97**, 2075–2080.

[53]   Hasty, J., F. Isaacs, M. Dolnik, D. McMillen, & J. J. Collins. 2001. Desinger gene networks: towards fundamental cellular control. *CHAOS* **11**, 207–220.

[54]   Gardiner, C. W. 1985. *Handbook of Stochastic Methods for Physics, Chemistry and the Natural Sciences*. Berlin: Springer-Verlag.

[55]   van Kampen, N. G. 1992. *Stochastic Processes in Chemistry and Physics*. Amsterdam: North-Holland.

[56]   Gillespie, D. T. 1977. Exact stochastic simulation of coupled chemical reactions *J. Phys. Chem.* **81**, 2340–2361.

[57]   Arkin, A., J. Ross, & H. H. McAdams. 1998. Stochastic kinetic analysis of developmental pathway bifurcation in phage λ-infected *Escherichia coli* cells. *Genetics* **149**, 1633–1649.

[58]   Kepler, T. B. & T. C. Elston. 2001. Stochasticity in transcriptional regulation: origins, consequences, and mathematical representations. *Biophys. J.* **81**, 3116–3136.

[59]   Mattis, D. C. & M. L. Glasser. 1998. The uses of quantum field theory in diffusion-limited reactions. *Rev. Mod. Phys.* **70**, 979–1001.

[60]   Eyink, G. L. 1996. Action principle in nonequilibrium statistical dynamics. *Phys. Rev. E* **54**, 3419–3435.

[61]   Gardner, T. S., C. R. Cantor, & J. J. Collins. 2000. Construction of a genetic toggle switch in *Escherichia coli*. *Nature* **403**, 339–342.

[62]   Aurell, E. & K. Sneppen. 2002. Epigenetics as a first exit problem. *Phys. Rev. Lett.* **88**, 048 101.

[63]   Walczak, A. M., J. N. Onuchic, & P. G. Wolynes. 2005. Absolute rate theories of epigenetic stability. *Proc. Natl Acad. Sci. USA* **102**, 18 926–18 931.

[64]   You, L., R. S. Cox III, R. Weiss, & F. H. Arnold. 2004. Programmed population control by cell-cell communication and regulated killing. *Nature* **428**, 868–871.

[65]   Guido, N. J., X. Wang, D. Adalsteinsson, *et al.* 2006. A bottom-up approach to gene regulation. *Nature* **439**, 856–860.

[66]   Guet, C. C., M. B. Elowitz, W. Hsing, & S. Leibler. 2002. Combinatorial synthesis of genetic networks. *Science* **296**, 1466–1470.

[67]   Kobayashi, M., C. Iaccarino, A. Saiardi, *et al.* 2004. Simultaneous absence of dopamine D1 and D2 receptor-mediated signaling is lethal in mice. *Proc. Natl Acad. Sci. USA* **101**, 11 465–11 470.

[68]   Fisher, R. A. 1930. *The Genetical Theory of Natural Selection*. Oxford: Clarendon Press.

[69]   Wright, S. 1932. The roles of mutation, inbreeding, crossbreeding and selection in evolution. *Proc. 6th Internat. Congress Genetics*, vol. 1, 356–366.

[70]   Delbruck, M. 1949. Discussion. In *Unités Biologiques Douées de Continuité Génétique Colloques Internationaux du Centre National de la Recheche Scientifique*. Paris: CNRS.

[71]   Waddington, C. H. 1957. *Strategy of the Gene*. London: Allen & Unwin.

[72]   Ao, P. 2005. Laws in Darwinian evolutionary theory. *Phys. Life Rev.* **2**, 117–156.

5

Modeling gene regulatory networks for cell fate specification

ARYEH WARMFLASH AND AARON R. DINNER

Introduction

Many cellular systems exist in more than one state. Two of the simplest and best-characterized examples are bacteriophage λ, which can be in the lytic or lysogenic state (Ptashne, 1992), and the *lac* operon of *Escherichia coli* bacteria which can be in either an active state, in which lactose is metabolized, or a repressed state, in which it is not (Novick & Weiner, 1957; Ozbudak *et al.*, 2004). These states can be remarkably stable. For example, in the absence of signals which induce the lytic cell fate, the likelihood of any given cell infected with phage λ switching from the lysogenic state is about 10^{-7} per generation (Aurell *et al.*, 2002). In multicellular organisms, a progenitor cell often selects between terminal cell fates which are even more stable; under natural circumstances, these decisions are irreversible. One complex but relatively well-studied example is the development of the cells of the immune system from a hematopoietic stem cell. The process of lineage commitment in this case involves a hierarchy of binary cell fate decisions in which a cell moves progressively closer to adopting certain terminal cell fates and loses access to others (Orkin, 2000; Medina & Singh, 2005).

Within an organism, almost all cells are genetically identical, but those in different states express different sets of proteins in order to perform specific functions. Which proteins a cell expresses is largely determined by the expression of transcription factors that regulate corresponding genes. In addition

Statistical Mechanics of Cellular Systems and Processes, ed. Muhammad H. Zaman. Published by Cambridge University Press. © Cambridge University Press 2009.

to controlling the expression of proteins that enable the cell to perform its functions, transcription factors also govern each others' expression in a complex network of regulatory interactions. There are two general possibilities for how such networks could control cell fate choice. One is that external regulatory signals induce expression of certain combinations of transcription factors to bias the regulatory dynamics towards a particular state. In this instructive model, cells adopt different fates due to exposure to different signals from their external environments. Alternatively, decisions could be stochastic: small fluctuations in the concentration of transcription factors become amplified to give rise to a stable cell state. These two mechanisms are not mutually exclusive and it is likely that both play a role in many cell fate decisions.

While the idea that the attractors of the dynamics of gene regulatory networks correspond to the stable cell fates observed in an organism was introduced many years ago (Kauffman, 1969; Glass & Kauffman, 1973), specific molecular realizations are just now beginning to be elucidated due to advances in both theory and experiment. In particular, the introduction of methods to measure protein expression in individual cells (Elowitz et al., 2002; Pedraza & van Oudenaarden, 2005; Rosenfield et al., 2005; Guido et al., 2006; Mettetal et al., 2006; Kaufmann et al., 2007) has driven interest in developing numerical (Allen et al., 2005; Erban et al., 2006; Giardina et al., 2006; Warmflash et al., 2007) and analytical (Thattai & van Oudenaarden, 2001; Paulsson, 2004; Warmflash & Dinner, 2008) approaches for treating stochastic fluctuations. Excitingly, the resulting dialog between modeling and single-cell experiments has already revealed cases in unicellular organisms in which fluctuations have a qualitative, rather than merely a quantitative, effect (Acar et al., 2005; Suel et al., 2006; Suel et al., 2007).

While a number of prototypical cell fate decisions in relatively simple organisms are now reasonably well understood through a combination of systems modeling and experiments (Ozbudak et al., 2004; Acar et al., 2005; Suel et al., 2006), lineage specification in more complex organisms is at best understood qualitatively, if at all. In the latter case, many of the key molecular players are now being identified, but turning individual nodes and connections in a regulatory network into a systems-level model capable of making reliable predictions is challenging because such gene regulatory networks often involve more than a dozen transcription factors (Orkin, 2000; Davidson et al., 2002; Medina & Singh, 2005; Rothenberg & Taghon, 2005). Consequently, computational models are likely to be even more important in understanding how quantitative changes in the expression levels of regulatory proteins lead to resolution of cell fates in multicellular organisms than unicellular ones.

Here, we discuss insights into cell fate decisions from modeling. We focus on the development of the mammalian hematopoetic system because it is relatively well studied by the standards of research on multicellular organisms and provides examples of a wide variety of cell fate decisions. Where appropriate, we compare and contrast the relevant regulatory network modules with better-characterized model systems. To this end, we first review deterministic methods used in computational studies of gene regulatory networks and apply them to elemental models of bi-stability. We show how these models can help elucidate many of the basic features of cell fate decisions. We then discuss extensions of this model applicable to specific cell fate decisions and show how they can be used to understand counterintuitive experimental results in complex systems. Next, we review stochastic methods for simulating gene regulatory networks and discuss their application to a simple model of a bi-stable switch. Finally, we discuss cell fate decisions with more complex dynamics and show how insights can be gained into these situations from a similar modeling framework.

Deterministic methods

Transcription factors regulate the expression of genes by binding to specific sites on the DNA which are typically spatially close to, sometimes even in, sequences that code for proteins. A group of such binding sites is collectively known as a *cis*-regulatory region. Once at their sites, the transcription factors act by influencing the rates of transcription of the regulated gene. When a gene processes the effects of multiple transcription factors, one can view it as performing a computation. The inputs are the occupancies of the binding sites in the *cis*-regulatory region and the output is the rate of transcription. The dependence of the rate of transcription on the concentrations of the inputs is often referred to as a "*cis*-regulatory input function (CRIF)." Such functions have been evaluated experimentally for the *lac* operon, which shows a complex dependence on the levels of the active forms of its transcription factors and for mutant versions which behave as pure AND and OR logic gates (Setty *et al.*, 2003; Mayo *et al.*, 2006).

What mathematical form should the CRIF take? Since generally transcription factors associate with and dissociate from DNA on a much faster timescale than transcription, translation, and degradation, binding and unbinding reactions can be assumed to be at equilibrium. In this approximation, the rate of transcription is determined by the average occupancy of the transcription factor binding sites at the current levels of the regulators. The CRIFs which result from this approximation are typically Michaelis–Menten-like or Hill functions of the transcription factor concentrations (Bintu *et al.*, 2005). For example, for a protein regulated by a single activator (A), the rate of transcription is given by

$$\text{CRIF}(A) = \frac{aA^{n_A}}{K^{n_A} + A^{n_A}}, \tag{5.1}$$

where a is the maximum rate of transcription due to the activator, K_A is the half saturation constant for activation (analogous to a Michaelis constant), and n_A is the Hill coefficient which accounts for cooperativity resulting from multiple binding sites or multimerization of the regulatory protein. Modeling repression as binding and inhibition of transcription, implicitly treating translation, and including protein degradation leads to a common general form for the time evolution of the concentration of a protein (x) (Elowitz & Leibler, 2000; Gardner et al., 2000; Angeli et al., 2004; Ozbudak et al., 2004; Acar et al., 2005; Laslo et al., 2006):

$$\frac{dx}{dt} = \left(\frac{aA^{n_A}}{K_A^{n_A} + A^{n_A}} + e_x \right) \frac{K_R^{n_R}}{K_R^{n_R} + R^{n_R}} - bx. \tag{5.2}$$

Here, R is the concentration of a repressor, n_R is the Hill coefficient for repression, K_R is the half saturation constant for repression, b is the rate of protein degradation, and e_x is the rate of protein synthesis in the absence of regulators. The term e_x enables one to account for inputs to the module of interest implicitly. If a gene is regulated by more than one activator or repressor, a more general form is required (see the Supplementary Information to Laslo et al., 2006 for discussion).

While common, (5.2) is not the only possible choice for the general form of the rate equations, and a form based on adding the contributions from activation and repression has also been employed (Huang et al., 2007). The additive model can be derived from an underlying one in which the effect of having both the activator and repressor bound is the sum of their individual effects, whereas the multiplicative model can be derived by assuming that when the repressor is bound, the gene is not transcribed even when the activator is present. When applied to the specific systems discussed below, the two different models share many qualitative features. More general models that do not make any assumptions about the relationship between the regulators of a gene can also be constructed; however, these require more parameters and there is little to be gained from assuming a more complicated model in the absence of experimental evidence to guide this choice.

Deterministic models, like ones defined by (5.2), neglect noise in protein concentrations and can therefore only be formally justified in cases in which the copy numbers are sufficiently large that their discrete and probabilistic nature can be neglected. Nonetheless, even in cases where this is clearly not true, the qualitative behavior of a system can often still be captured adequately. As such, deterministic models are often a valuable first step in gaining insight into the behavior of a complex network because mean-field models are typically far simpler to treat analytically and numerically than stochastic ones (reviewed below). In particular, there are well-defined means for determining the fixed points of the differential equations, which correspond to combinations of expression levels that do not change in time. Stable fixed points (or "attractors") are robust to perturbations and correspond to well-defined (but not necessarily mature) cell types. Unstable fixed points are not readily observable in experiments but are important because they define the boundaries (in expression levels) between developmental states. Once identified, one can assess the sensitivity of the fixed points to the parameters of the model, and trends can often be connected directly with experimentally observed behaviors or exploited to make predictions that can be used to validate the model experimentally.

Bi-stability

Since most, or perhaps all, cell fate decisions are binary in nature, gene regulatory network motifs that support bi-stability (i.e., have two attractors under the same conditions) are ubiquitous. Such motifs typically involve some form of positive feedback (Muzzey & van Oudenaarden, 2006). A key transcription factor can activate its own expression directly or indirectly through upregulating one of its activators (Fig. 5.1a). Alternatively, a transcription factor can antagonize one of its repressors (Fig. 5.2a). Functionally, the latter strategy ensures that lineage inappropriate genes are not misexpressed (Laslo *et al.*, 2006), whereas the former can help to maintain a stable cell fate in the absence of inducing signals (Xiong & Ferrell, 2003). In many transcriptional networks, both of these motifs are employed in tandem (Huang *et al.*, 2007), and we review each in turn.

Autoactivation

One of the simplest motifs giving rise to bi-stability is autoactivation, in which a protein stimulates transcription of its own gene (Fig. 5.1a). In this case, (5.2) takes the form

$$\frac{dx}{dt} = \frac{ax^{n_A}}{K_A^{n_A} + x^{n_A}} + e_x - bx. \tag{5.3}$$

This equation can be analyzed graphically by plotting the activation and degradation rates separately (Fig. 5.1b,c). When the two curves intersect, these processes are balanced and the corresponding concentration of x is a steady state (fixed point) of the model. The number of intersections therefore gives the number of fixed points of the model. In the case of cooperative ($n_A > 1$) and sufficiently

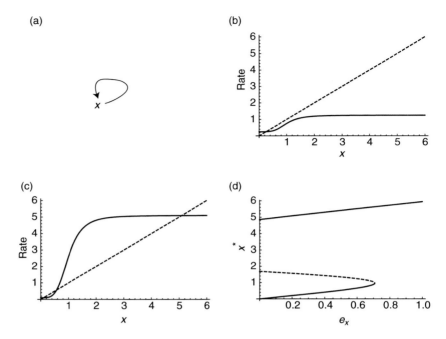

Fig. 5.1 Bi-stability resulting from autoactivation. (a) Schematic of autoactivation. (b and c) Graphical solution of the steady-state equations. Creation rates (solid lines) and degradation rates (dashed lines) plotted as functions of the concentration of x. The intersections of the lines represent solutions of the equations. (b) Standard parameters (see below) except $a = 1$ and $N_A = 4$. (c) standard parameters except $N_A = 4$. (d) Bifurcation diagram showing the dependence of the steady states on the level of external stimulus. Solid lines denote stable fixed points and dashed lines denote unstable fixed points. For all figures, $a = 5$, $b = 1$, $K = 1$, $n_A = 1$, and $n_R = 4$, unless otherwise indicated.

strong autoactivation, the model has three fixed points: two stable points in which x is either high or low (the gene is effectively either on or off) and an unstable point which separates the basins of attraction for these stable states.

As a result of the bi-stability, this module can "remember" a past stimulus (a transient increase in e_x or x itself in (5.3)) in that the stimulus can activate the positive feedback loop, which is then self-sustaining. Restricting attention to the case that the stimulus is represented by e_x, this behavior can be understood in terms of a bifurcation diagram, in which the steady-state values of x are plotted as a function of e_x. If the gene begins in the off state and the stimulus is increased, the gene will eventually be forced to transition to the on state at the point where the stable fixed point corresponding to low x no longer exists. If the stimulus is subsequently decreased, the cell can remain on the manifold

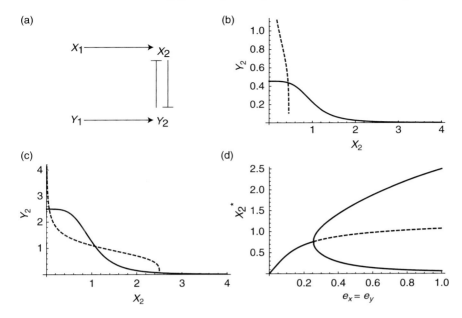

Fig. 5.2 Bi-stability resulting from mutual repression. (a) Schematic of mutual repression. Genes associated with one lineage are labeled X and those associated with the other lineage are labeled Y. The subscripts 1 and 2 denote primary and secondary cell fate determinants, respectively. (b and c) Solutions to the equations $dX_2/dt = 0$ (dashed lines) and $dY_2/dt = 0$ (solid lines) are shown. (b) $e_x = e_y = 0.1$. (C) $e_x = e_y = 1.0$. (d) Bifurcation diagrams showing the dependence of steady states on the rate of synthesis of the primary transcription factors. Parameters are as in Fig. 5.1 except as noted.

of stable fixed points corresponding to the on state even below the bifurcation (Fig. 5.1d). Thus, the state of the cell at $e_x = 0$ depends on its history: a gene that has not been exposed to a strong stimulus will be in the off state, and one that has will be in the on state. This form of memory is known as hysteresis and is a hallmark of bi-stable systems. A positive feedback loop composed of two mutually activating transcription factors exhibits analogous dynamics.

Mutual repression

Another motif which can govern a decision between two alternate cell fates is that of two mutually repressing transcription factors. To model a generic bi-stable switch which can be employed in studying cell fate decisions, we consider a circuit where each of the two transcription factors (X_2 and Y_2) is activated by a single upstream regulator (X_1 and Y_1) (Fig. 5.1b); we control the synthesis of X_1 and Y_1 through parameters like e_x in (5.2). This model has been employed in a variety of contexts (Gardner *et al.*, 2000; Laslo *et al.*, 2006). In the biological literature, the genes with subscripts 1 and 2 are called primary and secondary cell fate determinants, respectively (Singh *et al.*, 2005; Laslo *et al.*, 2006). Using one equation of the form of (5.2) for each gene, this network can be mathematically modeled as

$$\frac{dX_1}{dt} = e_X - bX_1 \qquad \frac{dX_2}{dt} = \frac{aX_1^{n_A}}{K_A^{n_A} + X_1^{n_A}} \frac{K_R^{n_R}}{K_R^{n_R} + Y_2^{n_R}} - bX_2 \qquad (5.4)$$

$$\frac{dY_1}{dt} = e_Y - bY_2 \qquad \frac{dY_2}{dt} = \frac{aY_1^{n_A}}{K_A^{n_A} + Y_1^{n_A}} \frac{K_R^{n_R}}{K_R^{n_R} + X_2^{n_R}} - bY_2$$

where the italicized X_i denotes the concentration of molecule X_i and for simplicity we have assumed that the degradation rates (b) and Michaelis constants ($K = K_A = K_R$) are the same for all genes. This allows us to choose the units of time such that $b = 1$ and the units of concentration such that $K = 1$. We generally assume that activation follows simple Michaelis–Menten-like kinetics ($n_A = 1$) and that repression is cooperative ($n_R > 1$) with the same n_R for all genes unless otherwise noted. We make this choice for activation because it is the simplest possible choice and for repression because cooperative repression is a necessary condition for bi-stability as shown below.

To begin to analyze (5.4), we consider the fixed points of the equations. By definition, these are the solutions to the algebraic equations obtained by setting the time derivatives to zero. The solution to the equations for X_1 and Y_1 is trivial ($X_1 = e_X$) so the problem reduces to solving:

$$X_2 = \frac{C_X}{1 + Y_2^{n_R}} \qquad Y_2 = \frac{C_Y}{1 + X_2^{n_R}} \qquad (5.5)$$

where $C_q = e_q^{n_A}/(1 + e_q^{n_A})$ for $q = X, Y$. The number and identity of the fixed points of the model depends on the number of physically meaningful solutions of (5.5). To solve these equations, it is helpful to plot the solution to each one separately as a line in the X_2Y_2-plane. The points where the lines intersect satisfy both of the equations and are therefore fixed points of the system. At low levels of e_X and e_Y, there is a single fixed point with nearly equal levels of X_2 and Y_2 (Fig. 5.2b), whereas at higher levels there are two additional fixed points, one with high X_2 and low Y_2 and one with the opposite configuration (Fig. 5.2c).

To proceed further, it is instructive to consider the symmetric case with $e_X = e_Y$ (so $C_X = C_Y = C$). Then symmetry dictates that the fixed point with equal levels of X_2 and Y_2 seen above is a solution of (5.5) with

$$X_2 = Y_2 = q_{fp} = \frac{C}{1 + q_{fp}^{n_R}} \qquad (5.6)$$

where q_{fp} is defined as the solution to the above equation. This fixed point does not correspond to a terminal cell fate but rather corresponds to equal expression of the genes for both lineages, a condition often found in progenitor cells. To examine its stability, one must examine the eigenvalues of the Jacobian evaluated at this fixed point. The fixed point is stable if both eigenvalues are negative (Strogatz, 1994). For our model, this corresponds to the condition

$$(n_R - 1)q_{fp}^{n_R} < 1. \tag{5.7}$$

In particular, for $n_R = 1$ this condition is always satisfied and this fixed point is stable. In fact, it is the only physically meaningful solution to (5.5). Thus, without cooperativity this system cannot function as a switch, and always responds in a graded fashion to increases in the levels of the activators. If $n_R > 1$, this solution is unstable for sufficiently large q_{fp}, or, equivalently, for sufficiently large e_X and e_Y.

To understand the behavior of the system when this mixed lineage fixed point is unstable, it is helpful to consider the case $n_R = 2$ in which case the other solutions to (5.5) can be determined analytically. Besides the mixed lineage fixed point, the other two fixed points satisfy

$$X_2 = \frac{C \pm \sqrt{C^2 - 4}}{2} \tag{5.8}$$

and by symmetry

$$Y_2 = \frac{C \mp \sqrt{C^2 - 4}}{2}. \tag{5.9}$$

These solutions are physically meaningful and stable if $C > 2$ which is the same point at which the mixed lineage fixed point becomes unstable. That is, for sufficiently large expression of the activators (sufficiently large e_X and e_Y) the mixed fixed point becomes unstable and the system differentiates to either a state with high X_2 and low Y_2 or one with the opposite configuration. These considerations are summarized in the pitchfork bifurcation diagram shown in Fig. 5.2d.

More generally, we can determine whether the system is in the mono-stable mixed lineage regime or in the bi-stable differentiated regime as a function of the rates of production of the activators (e_X and e_Y). The results can be used to draw a "phase" diagram for the system (Fig. 5.3a). In the model, the mixed lineage point is stable when the levels of the activators are low. This is in agreement with experiments which have revealed that progenitors promiscuously express proteins from a variety of lineages at low levels ((known as "transcriptional priming": Hu et al., 1997; Miyamoto et al., 2002; Laslo et al., 2006).

The model also provides a framework for uniting both the stochastic and instructive mechanisms for cell fate decisions. If the progenitor cell is destabilized with nearly equal concentrations of X_2 and Y_2 then stochastic fluctuations will determine which fate is ultimately adopted by the cell (dashed line, Fig. 5.3a). On the other hand, if e_X is increased before e_Y, then the state will remain in the X lineage even if e_Y is increased (solid line, Fig. 5.3a). Thus a differentiation signal which increased e_X prior to e_Y would represent an instructive

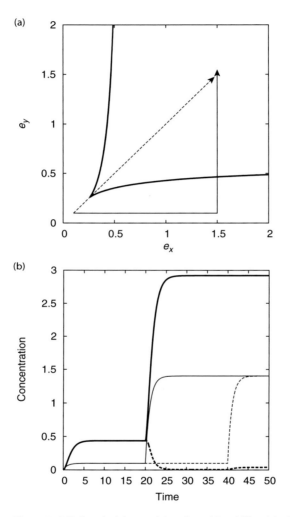

Fig. 5.3 Cell fate decisions arising from bi-stability. (a) The plot shows the border between the bi-stable and mono-stable regions as a function of e_X and e_Y. Examples of instructive (solid line) and stochastic (dashed line) pathways are shown. (b) The concentrations of the key regulatory proteins as a function of time as a cell moves along the instructive pathway shown in (a). Shown are curves for X_1 (thin solid line), X_2 (thick solid line), Y_1 (thin dashed line), and Y_2 (thick dashed line). At first, $e_X = e_Y = 0.1$. At time $t = 20$, e_X is increased to 1.5 and at time $t = 40$, e_Y is increased to 1.5. Other parameter values are as in Fig. 5.1.

pathway. Concentrations as a function of time along this instructive pathway are shown in Fig. 5.3b. This is another example of hysteresis. Transition to a state with high Y_2 following an increase in e_X requires that e_Y be raised to much higher levels than would have been necessary prior to this increase.

Application to cell fate decisions
Natural lineage specification

Variations on the basic models described in the previous section can be used to describe particular binary cell fate decisions. The autoactivation model is most applicable to situations in which a cell must "remember" exposure to a stimulus after that stimulus is no longer present. This model has been used to explain how T cells maintain their cell fate choice after transient exposure to an instructive signal. In this case, the autoactivating protein is GATA-3 and the instructive signal that controls e_x is the cytokine IL-4 (Hofer et al., 2002). If IL-4 is present during activation, cells upregulate GATA-3 and adopt the Th2 lineage rather than the Th1 lineage. The former is then stable even in the absence of IL-4. Autoactivation also plays a role in the decision between the erythroid and myeloid lineages. Here it helps ensure the stability of these lineages following the cell fate decision which is governed by mutual repression (Huang et al., 2007) (discussed further below). The same regulatory motif has been used to understand maturation of Xenopus oocytes whose development is governed by coupled positive feedback loops in a signaling cascade (Xiong & Ferrell, 2003) (Fig. 5.4a). The oocytes begin in the immature state and, depending on the presence of the stimulus, either transition to the mature state or do not. Those that have transitioned remain in the mature state even in the absence of the stimulus due to coupled positive feedback loops, each of which can be described by (5.3) (Xiong & Ferrell, 2003). The existence of multiple loops with varying timescales has been suggested to allow both a sensitive response due to the fast loops and a robust one due to the slow loops (Brandman et al., 2005). Qualitatively similar models have been applied to the decisions made by E. coli and yeast on whether to transition from the (default) state in which a nutrient is not metabolized to one in which it is (Ozbudak et al., 2004; Acar et al., 2005).

The mutual repression model is applicable to a wide variety of binary decisions made between mutually exclusive cell states. In the hematopoietic system, different types of myeloid cells (macrophages and neutrophils) develop from a common progenitor. The primary and secondary cell fate determinants for the macrophage fate are PU.1 and Egr-2, respectively, while those for the neutrophil lineage are C/EBPα and Gfi-1 (Fig. 5.4b). The main difference between this network and the basic one described above is that the secondary determinant for the neutrophil lineage (Gfi-1) is thought to repress the primary cell

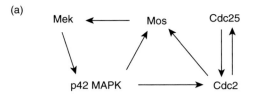

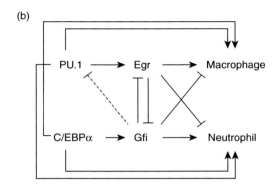

Fig. 5.4 Network diagrams for cell fate decisions. (a) *Xenopus* oocyte developmental signaling network. (b) Macrophage–neutrophil network.

fate determinant for the other (PU.1) (Laslo *et al.*, 2006; Dahl *et al.*, 2007). This interaction breaks the symmetry in the network and complicates the analysis, but the dynamics are qualitatively similar to those described above (Laslo *et al.*, 2006).

The regulation of the expression of a macrophage-specific gene (*c-fms*) downstream of PU.1 and Egr-2 has been described in detail (Krysinska *et al.*, 2007) and can therefore serve as a model for how the primary and secondary cell fate determinants jointly regulate genes further downstream. The *c-fms* gene is first induced at low levels by the presence of PU.1 and later at higher levels by Egr-2. Interestingly, C/EBPα, the primary cell fate determinant for the neutrophil lineage, also serves as an activator for this macrophage gene. In fact, this observation is part of a larger paradox concerning myeloid development. Counterintuitively, differentiated cells express high levels of primary cell fate determinants from both lineages and binding sites for both of these proteins are found in the regulatory regions of lineage specific downstream genes (Laslo *et al.*, 2006). The model reveals that increasing the levels of both primary cell fate determinants destabilizes the mixed lineage progenitor and forces differentiation to one of the terminal cell fates. Thus, increasing the expression of both primary cell fate determinants can be used as a strategy for both lineage commitment and stochastic specification (Laslo *et al.*, 2006).

A similar model that also included positive autoregulation for each of the secondary cell fate determinants (X_2 and Y_2 activate their own production) was used to model an earlier developmental decision in the hematopoietic system, that between the myeloid and erythroid lineages (Huang *et al.*, 2007). The behavior in this case is qualitatively similar except that, in addition to the mono-stable and bi-stable regimes, there is a tri-stable regime in which the progenitor and both cell fates are stable. In this case, as in the previous one, the progenitor state is destabilized when expression of the primary cell fate determinants is sufficiently high, so the cell must transition to one cell fate or the other. The existence of the tri-stable regime means that the mixed lineage fixed point is destabilized at higher levels of the primary cell fate determinants than are necessary for the creation of the fixed points corresponding to the differentiated cell fates, whereas in the previous model for the macrophage–neutrophil decision these two events occurred at the same concentration of primary determinants. In nonlinear dynamics, these two situations are known as sub-critical and super-critical pitchfork bifurcations, respectively (Strogatz, 1994). In principle, these can be distinguished by the dynamics of the system in the X_2Y_2-plane, and some experimental support for the sub-critical case was provided by analysis of microarray data in the myeloid–erythroid case (Huang *et al.*, 2007).

For comparison, a bi-stable circuit understood in much greater detail is that which governs the lysis–lysogeny decision in bacteriophage λ (Ptashne, 1992). This can be viewed as a more complex version of the simpler switch shown in Fig. 5.6a (below). The core of this circuit is mutual repression between Cro, which triggers the lytic pathway, and cI, which maintains the lysogenic state. The λ genome contains two regulatory regions which govern the expression of these proteins (O_R and O_L) each of which contains three binding sites capable of binding either dimers of Cro or dimers of cI. Cooperativity in the regulation of cI and Cro stems from dimerization of these proteins as well as interactions between neighboring binding sites and long-range interactions between the O_R and O_L regions resulting from DNA looping (Dodd *et al.*, 2005). The stability of the lysogenic state is enhanced by positive autoregulation of the cI gene. Nonetheless, the virus is poised to trigger the lytic state in response to external stimuli because cI also represses its own expression when present at high levels. Because measurements of the parameters of almost all of these interactions have been made, a quantitative model can be assembled using similar methods to those outlined above and used to understand the influence of and interplay between these different interactions (Santillan & Mackey, 2004). The layers of complexity which serve to both guarantee the stability of the differentiated state and at the same time allow sensitive response to changes in environmental conditions in this simple organism indicate that, while cell fate decisions in mammalian systems are beginning to be qualitatively understood, it is likely that many additional interactions which serve to both induce these cell fate decisions and render them irreversible remain to be elucidated.

Understanding reprogramming experiments

Recently, there has been increased interest in the robustness and plasticity of cell fate decisions. As a cell progresses down a developmental pathway, what alternatives are available and when does commitment to a given cell fate become irreversible? These questions can be explicitly probed by "reprogramming" experiments in which the expression of a transcription factor known to promote the development of an alternate lineage is enforced (Klinken *et al.*, 1988; Kondo *et al.*, 2000; Xie *et al.*, 2004; Laiosa *et al.*, 2006; Cobaleda *et al.*, 2007; Taghon *et al.*, 2007). These experiments suggest that common myeloid and lymphoid progenitors (as defined by cell surface proteins) and even differentiated cells that arise from them are surprisingly plastic and can be readily made to produce alternate lineages.

Although most reprogramming experiments are clearly consistent with regulatory relationships established by independent means (Klinken *et al.*, 1988; Kondo *et al.*, 2000; Cobaleda *et al.*, 2007), some yield counterintuitive results. In particular, Graf and co-workers showed that forced expression of the neutrophil promoting transcription factor (C/EBPα) in B and T lymphocyte precursors converted these cells into macrophages rather than neutrophils (Xie *et al.*, 2004; Laiosa *et al.*, 2006). To understand these reprogramming experiments, we expanded the myeloid network described above (Laslo *et al.*, 2006) to include key cell fate determinants for the B cell lineage (Seet *et al.*, 2004; Medina *et al.*, 2004; Dias *et al.*, 2005; Kikuchi *et al.*, 2005; Singh *et al.*, 2005; Singh *et al.*, 2007) (Fig. 5.5a). A similar analysis can be performed for the T cell case by including GATA-3, Notch, and HEB, but the qualitative conclusions are the same (data not shown).

The combined lymphoid–myeloid network in Fig. 5.5a has stable fixed points corresponding to the B cell, neutrophil, and macrophage cell fates. We simulated experiments in which C/EBPα was induced in a progenitor cell which would have adopted the B cell fate if left unperturbed. Parameters were set to those that produce B cells (Fig. 5.5b) and the rate of synthesis or concentration of C/EBPα was spiked at various points along this pathway. When the rate of C/EBPα synthesis was spiked very early in development, neutrophils resulted (Fig. 5.5c). However, if the cell was allowed to proceed along the B cell pathway, spiking the C/EBPα rate of synthesis produced macrophages (Fig. 5.5d). In the first case, both Egr-2 and Gfi-1 were at low concentrations when the spike occurred, such that C/EBPα promoted the neutrophil pathway at the expense of the macrophage

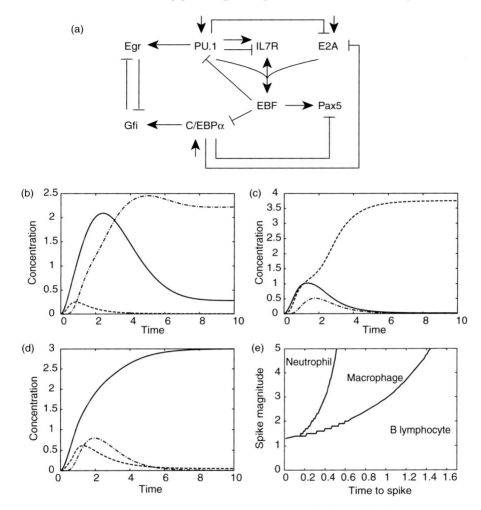

Fig. 5.5 Reprogramming experiments. (a) Schematic of combined lymphoid–myeloid network used for reprogramming simulations. Small arrows indicate genes with a nonzero rate of default synthesis ($e_x \neq 0$). (b through d) Time series for key cell fate determinants. The concentrations shown are for the B lineage gene EBF (dashed-dotted lines), the macrophage gene Egr (solid lines), and the neutrophil gene Gfi-1 (dashed lines). (b) Natural case. Time course for B cell development without reprogramming. (c and d) The rate of C/EBPα synthesis is spiked by a unit amount at time (c) $t = 0.3$ or (d) $t = 0.6$. (e) "Phase" diagram for reprogramming. The final cell fate is shown as a function of the timing and magnitude of the spike in C/EBPα synthesis. Parameters were the same as in Fig. 5.1 except that $\alpha_{EBF} = 3$ because higher values of this parameter made reprogramming impossible. EBF induction requires integrating signals from PU.1, IL7R, and E2A. To reflect this fact, we made the substitution $A^{n_A} \rightarrow [\text{PU.1}][\text{IL7R}]^2[\text{E2A}]^2$ where the choice of Hill exponents is such that EBF is activated only by low levels of PU.1. The results are not sensitive to the specific choice of exponents as long as those for IL-7R and E2A are larger than that for PU.1.

pathway. In the second case, since Egr-2 had been activated by PU.1, which is obligate for lymphoid as well as myeloid development, subsequent activation of C/EBPα could not activate Gfi-1. However, since C/EBPα represses several B cell specific genes, it allowed cells to default into macrophages.

These results are summarized in Fig. 5.5e. Early induction of C/EBPα in sufficient quantities reprograms lymphocytes to neutrophils, and later induction reprograms them to macrophages. However, as cells develop along the lymphocyte pathways, it becomes increasingly difficult to divert cells to myeloid fates. In considering these results, it is important to note that Gfi-1 is expressed in lymphocytes (Hock & Orkin, 2006) . However, the model only requires that the levels of Egr be sufficiently high that the Egr–Gfi cross-antagonism gives rise to a bi-stable switch biased to the macrophage-promoting side immediately following the spike in C/EBPα expression.

In summary, modeling reveals that lymphocyte development biases the bi-stable switch to the macrophage-promoting side, so that induction of myeloid genes (even neutrophil-promoting ones) can force the cell to adopt a macrophage fate. This mechanism is consistent with the finding that PU.1 is necessary to reprogram lymphocyte progenitors to macrophages but not to downregulate lymphocyte-specific genes (Xie *et al.*, 2004; Laiosa *et al.*, 2006). C/EBPα itself causes the downregulation of these genes, but PU.1 is necessary to bias the switch to the macrophage side. The results of the modeling could be tested by performing similar reprogramming experiments in which the levels of Egr-1,2 or Gfi-1 were perturbed either through genetic knockouts or enforced expression. For example, in the case in which Gfi-1 expression was enforced, we would expect the switch to be biased to the neutrophil side and C/EBPα to reprogram B cell progenitors to neutrophils.

Stochastic models and their treatment

Recent advances in methods for measuring mRNA and protein expression in single cells have driven a dramatic expansion of interest in stochastic models of gene regulatory networks. Fluorescent proteins can be engineered into cells and their expression levels quantitated by flow cytometry or microscopy. The former method has the advantages that very large numbers of cells can be examined and it can be applied in a high-throughput manner to a large number of genes (Bar-Even *et al.*, 2006; Newman *et al.*, 2006). Microscopy complements flow cytometry in that it enables one to track the actual dynamics of a relatively small number of cells over time. This feature can be particularly important in the study of transient differentiation events (Suel *et al.*, 2006; Suel *et al.*, 2007). Sophisticated fluorescence-based techniques are now even capable of measuring protein copy numbers at the single-molecule level (Cai *et al.*, 2006; Yu *et al.*, 2006) and can be used to observe the dynamics of transcription factor binding to DNA (Elf *et al.*, 2007). Flow cytometry and time-lapse microscopy are not limited to unicellular organisms and both are in common use in more complex systems (Kallies *et al.*, 2004; Sigal *et al.*, 2006; Crouch *et al.*, 2007).

These experiments provide examples in which intrinsic noise in reactions involving low copy numbers of molecules can have important biological conse-quences (Elowitz *et al.*, 2002; Vilar *et al.*, 2002; Pedraza & van Oudenaarden, 2005; Suel *et al.*, 2006). Stochastic effects are thought to play an important role in cell fate specification (Laslo *et al.*, 2006) and have been shown to mediate transitions between two stable states (Acar *et al.*, 2005; Suel *et al.*, 2006; Kaufmann *et al.*, 2007). Although deterministic models can be suitable for determining the attrac-tors of a system, the robustness of the stable fixed points to noise and the transi-tion rates between them can only be determined from stochastic analyses. When sufficient numbers of cells are examined, the distribution of protein copy num-bers can be compared with the predictions of stochastic models; this has been done in simple bacterial systems and the distributions from theory and exper-iment agree well (Guido *et al.*, 2006; Mettetal *et al.*, 2006). These studies provide a proof of concept that stochastic models at the level of chemical kinetics can reproduce the distribution of protein copy numbers with quantitative accuracy.

The chemical master equation, which describes the time evolution of the joint probability distribution for the copy numbers of all species in the system, provides a rigorous framework for treating stochastic fluctuations in biological regulatory networks (van Kampen, 1992). In general, such an equation can be written as

$$\dot{P}_i = -P_i \sum_j r_{ji} + \sum_j r_{ij} P_j \qquad (5.10)$$

where i and j label sets of copy numbers for all the species, P_i is the probability of the system residing in state i, and r_{ij} is the rate of transition from state j to state i. The first and second terms account for transfer of probability out of and into state i, respectively. The complete master equation consists of a set of equations of the form of (5.10), one for each state in the system. Note that r_{ij} is usually not a constant but rather a function of the state of the system. This equation can be difficult to treat because most open systems can exist in an infinite number of states and r_{ij} is often a nonlinear function of the copy numbers of the molecular species. In practice, the master equation can be solved exactly only in linear systems (Gadgil *et al.*, 2005; Heuett & Qian, 2006; Jahnke & Huisinga, 2007) or in a handful of specialized cases (Hornos *et al.*, 2005).

It is still possible to proceed analytically using expansion methods such as the Kramers–Moyal and Ω expansions (van Kampen, 1992). The latter is formally justifiable because it involves a systematic expansion in a small parameter (Ω^{-1}), which can typically be chosen to be the inverse of the volume of the system. The lowest-order terms in the expansion provide a rigorous means for deriving the deterministic chemical rate equations from the master equation, and higher-order terms determine the statistical properties of the noise. The moments of the distribution can be calculated systematically to any order in Ω^{-1}. This approach has been used to derive a noise summation rule (Paulsson, 2004) and has been applied to specific models of gene regulatory (Thattai & van Oudenaarden, 2001) and signaling (Tanase-Nicola *et al.*, 2006) networks.

We recently used the Ω expansion to derive a class of fluctuation–dissipation-like relations for a simple model of DNA logic which integrates information from an arbitrary number of transcription factors (Warmflash & Dinner, 2008). As discussed above, the input–output relationship of a gene can be described by a CRIF which gives the rate of transcription as a function of the concentrations of inputs. If one is interested in how two transcription factors jointly regulate the output of a gene, the second mixed partial derivative is of central importance because it contains information about how the response of the gene to one of the regulators is influenced by the presence of the other. Using the Ω expansion, we find:

$$\frac{\partial^2}{\partial X_i \partial X_j} \text{CRIF}(X_1, \dots, X_n) = C\langle \delta X_i \delta X_j \delta R \rangle \tag{5.11}$$

where X_i and X_j are two of the n inputs to the gene, R is the output of the gene, C is a positive constant, and $\delta X_i = X_i - \langle X_i \rangle$ is the difference between the measured concentration of X_i in a single cell and its average across the population. This relation is general and does not depend on any specific form of the CRIF. Such a relation is useful because it connects a quantity which can be calculated by simple arithmetic manipulation of data from a straightforward flow cytometry experiment with the central quantity that describes how the gene responds to X_i and X_j. Since C is positive, these two quantities have the same sign, so that one can robustly determine the curvature of the CRIF, which relates directly to whether the gene performs logic more like an AND operation or more like an OR operation. Thus the Ω expansion enables one to obtain quite general relations for inferring information about combinatorial control of transcription from single-cell measurements. The essential idea is that natural fluctuations can substitute for artificial manipulations of the regulators.

Unfortunately, the Ω expansion breaks down at points in which the deterministic rate equations do not flow to a single stable fixed point, such as on the dividing line between two attractors, because it assumes that the fluctuating part of the solution is smaller than the deterministic part by a factor of

$\Omega^{-\frac{1}{2}}$, which is not the case at these points. While it can serve as a basis for mapping gene regulatory interactions (Warmflash & Dinner, 2008), it is thus of limited applicability in analyzing the cell fate decisions themselves. The eikonal expansion (Dykman et al., 1994) is similar in spirit in that it is also based on an expansion in the inverse volume, but it does not involve any assumptions about the relative magnitude of the (retained) terms in the expansion and therefore does not suffer from this difficulty. The eikonal expansion has been used to calculate most probable paths and switching rates for a simple stochastic model of a bi-stable switch similar to those reviewed below (Roma et al., 2005).

While the above analytical methods are useful for analyzing small systems with only a few components and interactions, they quickly become intractable for more complicated problems. Likewise, numerical approaches based on determining the time evolution or steady state of the probability distribution by directly solving the master equation are also prohibitively costly for large systems. These difficulties can be overcome by simulating trajectories consistent with the master equation (Gillespie, 1977). Accumulating statistics over a large number of trajectories allows the probability distribution and related averages to be determined. The basic idea is that, despite the fact that it is not possible to solve the master equation, the probability distribution for the next reaction and the time to that reaction are known exactly starting from any given state. First, the propensity (the rate constant multiplied by the concentrations of reactants) for each reaction is computed; we denote the propensity for reaction j by $a_j(t)$ and the sum of all propensities by $a_{tot}(t) = \sum_j a_j(t)$. $a_{tot}(t)$ gives the total rate to exit from the current state and is a constant for any given state. Because this rate is independent of the amount of time spent in the current state, the exit process is a Poisson process and the probability to remain in the current state for a time Δt falls off exponentially with time and the total propensity: $P(\Delta t) = a_{tot}e^{-a_{tot}\Delta t}$. In practice, this distribution can be sampled by choosing the time until the next reaction as:

$$\Delta t = -\frac{\ln r_1}{a_{tot}} \tag{5.12}$$

where r_1 is a random number chosen from a uniform distribution on the interval $0 < r_1 \leq 1$. Note that in contrast to traditional numerical integration methods where a constant time step is employed, here, each step corresponds to a discrete event and the time counter is advanced accordingly. The probability that reaction m is the next reaction is proportional to $a_m(t)$, so a second random number (r_2) is chosen and the reaction actually executed is the one with index m such that:

$$\sum_{j=1}^{m-1} a_j(t) < r_2 a_{tot}(t) \leq \sum_{j=1}^{m} a_j(t). \tag{5.13}$$

Equivalent schemes that reduce the number of random numbers needed and are thus computationally more efficient have been introduced (Gibson & Bruck, 2000). Other details can also influence the efficiency of the algorithm and have been reviewed elsewhere (Gillespie, 2007).

Even the most efficient such algorithms can be prohibitively costly to run in systems which have a large separation of timescales between relevant processes. This is often the case for biological problems such as cell fate determination because the time to transition between stable states is much longer than the timescales associated with the individual reactions of DNA binding, protein production, and degradation. Furthermore, even these reactions themselves can exhibit a wide range of timescales with DNA binding and chemical modifications such as phosphorylation occurring on a much faster timescale than transcription, translation, and protein degradation (Guido *et al.*, 2006). In these cases, most of the computational time is spent executing the fast reactions while switching between cell fates occurs rarely if at all. One strategy for overcoming this problem is to introduce approximations. There are essentially four classes of algorithms in this regard: (1) ones that treat the evolution of some of the components by deterministic equations (Takahashi *et al.*, 2004; Salis & Kaznessis, 2005a; Wylie *et al.*, 2006), (2) ones that assume that some of the reactions are effectively at equilibrium (Salis & Kaznessis, 2005b), (3) ones that assume a separation of timescales and execute fast reactions repeatedly in bursts as in leaping methods (Gillespie, 2001; Rathinam *et al.*, 2003) (reviewed in (Gillespie, 2007)), and (4) ones that employ many short simulations from a number of different initial conditions and then combine the information from these into a coarse-grained model (Erban *et al.*, 2006; Shalloway & Faradjian, 2006).

However, there are now efficient alternatives that do not rely on approximations. These algorithms are essentially extensions of methods introduced for sampling rare events in reversible systems (Valleau & Card, 1972; Dellago *et al.*, 1998; Bolhuis *et al.*, 2002; Frenkel & Smit, 2002; van Erp *et al.*, 2003). What made the development of like methods that are suitable for treating gene regulatory networks challenging is that such systems do not obey detailed balance. As such, statistics for a region of phase space can be determined by a balance of flows through different possible transitions. In nonequilibrium umbrella sampling (Warmflash *et al.*, 2007), equal sampling of different regions of a space of an arbitrary number of order parameters is enforced by restricting copies of the system to limited ranges of the order parameters with careful accounting for the fluxes in and out of those ranges. In forward flux sampling (Allen *et al.*, 2005; Allen *et al.*, 2006), the space between two attractors is divided by a series of nonintersecting interfaces, and trajectories connecting sequential interfaces are used to construct full transition paths and estimate transition rates between

Table 5.1. *Reactions for the genetic toggle switch. The rate constants for the forward and backward reactions are* k_f *and* k_b; "..." *indicates that there is no backward reaction;* Ø *denotes degradation. The parameters for the dimerization reactions are a factor of two larger than those reported by Allen* et al. *(2005) because otherwise it was not possible to reproduce their results when correctly accounting for the indistinguishability of the reactants (Gillespie, 1977).*

A reactions	B reactions	k_f	k_b
$A + A \rightleftharpoons A_2$	$B + B \rightleftharpoons B_2$	10	5
$O + A_2 \rightleftharpoons OA_2$	$O + B_2 \rightleftharpoons OB_2$	5	1
$O \rightarrow O + A$	$O \rightarrow O + B$	1	...
$OA_2 \rightarrow OA_2 + A$	$OB_2 \rightarrow OB_2 + B$	1	...
$A \rightarrow \emptyset$	$B \rightarrow \emptyset$	0.25	...

the attractors. Both of these algorithms have been combined with the Gillespie algorithm (Gillespie, 1977) outlined above to simulate gene regulatory networks efficiently.

Stochastic description of a bi-stable switch

We now consider a simple stochastic description of a bi-stable switch. The chemical reactions which define the system are given in Table 5.1. Two proteins, A and B, can each homodimerize and then bind to an operon(O). The operon can only bind one dimer at a time. When a dimer of A (B) is bound to the operon, it represses transcription of the gene for B (A); there is no concomitant effect on the expression of A (B). As a result of these dynamics, the system has two stable states: one with abundant A and scarce B and the opposite situation (Fig. 5.6). Switching between the two stable states is rare: approximately six orders of magnitude less frequent than each elementary reaction (Allen et al., 2005). This model has been investigated in a large number of studies both because of its intrinsic interest as a simplified model of a biological system (Warren & ten Wolde, 2004; Warren & ten Wolde, 2005; Lipshtat et al., 2006) and as a test case for new algorithms for sampling nonequilibrium processes (Allen et al., 2005; Erban et al., 2006; Valeriani et al., 2007; Warmflash et al., 2007).

Analysis of this model has helped elucidate several important features of bi-stable switches. These are as follows. (1) The paths traversed when traveling from the stable state with high A to the one with high B are not the same as those traversed when traveling in the opposite direction (Allen et al., 2005). This is a consequence of the fact that the dynamics of the model violate detailed

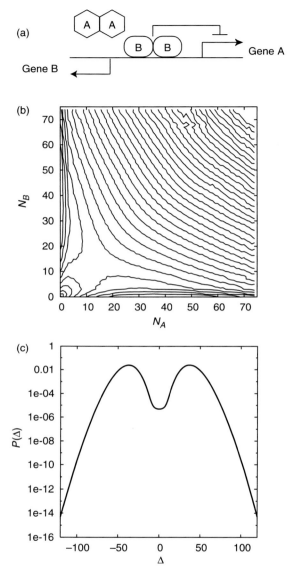

Fig. 5.6 Stochastic description of a genetic toggle switch. (a) Schematic of the toggle switch. (b) Steady-state probability distribution for the toggle switch computed with nonequilibrium umbrella sampling (Warmflash *et al.*, 2007). Contours are logarithmically spaced by factors of 10 with the innermost and outermost contours corresponding to probabilities of 10^{-3} and 10^{-35}, respectively. N_A and N_B are the total numbers of A and B molecules. (c) Probability distribution along the variable $\Delta = N_A - N_B$. The system is approximately five orders of magnitude less likely to be in the transition region than in either of the stable states. See (Warmflash *et al.*, 2007) for details.

balance. (2) The fact that the operon in the model can only bind one dimer at a time greatly increases the stability of the switch. Simulations of a model in which both proteins could bind simultaneously showed greatly decreased times between transitions (Warren & ten Wolde, 2004). (3) The stability of the peaks can be enhanced by increasing the copy numbers of the proteins expressed in the stable states. Simulations have shown that in this model the lifetime of the stable states grows with the copy number of proteins as $\tau \sim N^a e^{bN}$ where a and b are constants (Warren & ten Wolde, 2005). (4) As a result of this scaling, the relative stabilities of the stable states can be very sensitive to the values of the parameters. Increasing the degradation rate for one of the proteins by a factor of two makes the stable state with high levels of that protein approximately five orders of magnitude less likely to be occupied than the stable state with high levels of the other (Warmflash et al., 2007). (5) While cooperativity is required to produce bi-stability in the deterministic description, a variant of this model without dimerization contains no cooperativity but nevertheless exhibits switch-like behavior (Lipshtat et al., 2006).

Finally, it is worth noting that the deterministic limit of this model is essentially that outlined for mutual repression above (5.4) with Hill coefficients of $n_R = 2$. Here, $n_R = 2$ stems from the homodimerization. This is not the only possible mechanism that gives rise to those kinetics in the mean field. Another such model would be one in which the gene had two binding sites for the protein and full repression occurred only when both binding sites were occupied. Alternatively, one could simply start from a phenomenological perspective and let the creation rate be a nonlinear function as in the first term on the right-hand side of (5.2). We mention these different scenarios to make the point that, while they all have the same deterministic rate laws, each has different intrinsic noise properties. For this reason, it can be of interest to include as much detail as is computationally feasible in the stochastic case. On the other hand, the deterministic fixed points of all the models and thus their gross qualitative behavior are identical, so there is little to be gained by eschewing the more phenomenological model in the absence of compelling experimental support for one mechanism or another.

Even the relatively detailed model defined by the reactions in Table 5.1 is still a considerably simplified description of a real biological system. For example, protein production is treated as a single step rather than including the details of transcription and translation. This choice also alters the noise properties since many biological systems are affected by noise associated with infrequent bursts of protein production that occur whenever an mRNA molecule is produced (Thattai & van Oudenaarden, 2001; Bar-Even et al., 2006; Warren et al.,

2006). As more becomes known about systems and more quantitative questions are asked about them, it will become important to include the details of DNA binding, transcription, and translation. As a result, simulation methods that enable one to overcome sampling problems associated with separations of timescales, like the various approximate and exact algorithms reviewed above, will become increasingly important in the future for treating phenomena of genuine biological interest.

Elaborations on bi-stability

Although the models of bi-stability discussed above provide a useful tool for examining cell fate decisions, not every cell fate decision can be viewed as a simple bi-stability. This is particularly true in cases in which the cell fate decision is intrinsically transient as opposed to one which is stable in the absence of external cues. In such cases, the situation is more complicated because the transient cell fate must have a range of attraction but at the same time be itself unstable. To understand how such a combination can arise, it is important to examine the dynamics as well as the fixed points of the model and to consider how stochastic effects influence the lifetime of the stable states.

An example of an intrinsically transient differentiation process is presented by mature B lymphocytes following activation (Abbas *et al.*, 2000). If a B cell encounters antigen which is recognized by its receptor, it first enters the germinal center state in which it can switch the type of antibodies it produces (isotope switch) and mutate the variable portions of its antibodies to increase their affinity for the antigen (affinity mature). The cell then differentiates to either a memory B cell, which is poised to respond to future infections, or to an antibody-secreting plasma cell, which helps fight the current infection. One of the most interesting aspects of this process is that, while the germinal center and plasma cell states are mutually exclusive, the same transcription factor, IRF-4, appears to induce both (Sciammas *et al.*, 2006). Low to medium levels of IRF-4 promote the germinal center state, and high levels promote the plasma cell state. A minimal network governing this process is shown in Fig. 5.7a.

The approach defined by (5.2) can easily be used to model this network, and a full discussion of its biological implications will be published elsewhere. Here, we consider a toy model with the same basic physics as the full network (Fig. 5.7b). This network contains only IRF-4 (labeled I) and only one gene each for plasma cell and germinal center fates (B and C respectively). B and C repress each others' expression to form a bi-stable switch similar to that modeled above. I drives both sides of this switch asymmetrically: low levels of I bias the switch to the C side, while high levels bias it to the B side.

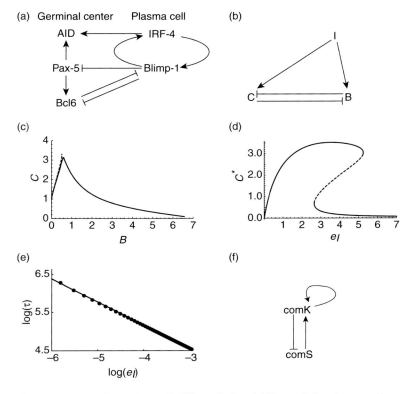

Fig. 5.7 Models of transient cell differentiation. (a) Network for plasma cell development. (b) Simplified model with the same basic physics as the network in part (a). (c) Trajectories in the BC-plane with the rate of IRF-4 synthesis below the bifurcation point (dashed line, $e_I = 5.0$) and above the bifurcation point (solid line, $e_I = 5.6$). (d) Bifurcation diagram showing the steady-state value of C as a function of the rate of IRF-4 synthesis. (e) Scaling behavior of time spent in the germinal center with distance in parameter space from the bifurcation point. Data points are the results of numerical simulations and the solid line is fit to the equation $\log \tau = \alpha + \beta \log \Delta e_I^*$ with $\alpha = 2.73$ and $\beta = 0.61$. For these simulations we choose $n_{A_B} = 2$ for the activation of B by I, which reflects the cooperativity due to the positive feedback loop between Blimp-1 and IRF-4 in the full network. All other parameters are the same as those in Fig. 5.1 except as noted in the text. (f) Schematic of the network governing competence induction.

Mathematically, in a deterministic description, the reduced network can be written as the following set of equations:

$$\frac{dI}{dt} = e_I - bI$$

$$\frac{dC}{dt} = \left(\frac{a_C I^{n_{AC}}}{K_C^{n_{AC}} + I^{n_{AC}}} \right) \frac{K_R^{n_R}}{K_R^{n_R} + B^{n_R}} - bC$$

$$\frac{dB}{dt} = \left(\frac{a_B I^{n_{AB}}}{K_B^{n_{AB}} + I^{n_{AB}}} \right) \frac{K_R^{n_R}}{K_R^{n_R} + C^{n_R}} - bB. \qquad (5.14)$$

We make the choices $K_C < K_B$ and $a_C < a_B$. The former ensures that the germinal center gene (C) is activated at lower levels than the plasma cell gene (B), whereas the latter ensures that the switch is biased towards the plasma cell side when IRF-4 is expressed at high levels. Analysis of this model shows that, at low levels of I induction, there are two stable states corresponding to the germinal center and plasma cell programs. Although both states are present, only the germinal center state is kinetically accessible because resting B cells express key germinal center determinants (such as Pax-5) and thus the switch is initially biased to the germinal center side. If I is produced at sufficiently high levels, the germinal center fixed point is lost and the plasma cell state becomes the only stable fixed point of the system (Fig. 5.7d).

One of the most interesting features of this transition is that even when I is raised to levels sufficient to induce plasma cell differentiation from the resting B cell state, the cell still initially moves towards the germinal center state. It approaches the point in the phase space where the germinal center fixed point was located at lower I before finally transitioning to the plasma cell state (Fig. 5.7c). This effect is known in the nonlinear dynamics literature as "ghosting" because the system feels the remnants of a fixed point even after it has been lost through a bifurcation (Strogatz, 1994). This feature of the model provides a possible explanation for the transient differentiation dynamics observed in these cells.

One of the hallmarks of ghosting is that the dynamics exhibit a slowing down near the ghost of the fixed point such that the time spent near this point scales with the parameter governing the bifurcation as

$$\tau \sim \Delta e_I^{-\beta} \qquad (5.15)$$

where $\Delta e_I = e_I - e_I^*$, e_I^* is the value of e_I at the bifurcation point, and the value of β is usually 0.5 due to the quadratic nature of equations of motion near the bifurcation point. Thus, the model predicts that, the closer the rate of IRF-4 is to the bifurcation point, the longer a cell will remain in the germinal center (Fig. 5.7e). Since there is evidence that IRF-4 is induced by signaling through the B cell receptor (R. Sciammas and H. Singh, unpublished) (Benson et al., 2007), this mechanism could allow cells with lower affinity for antigen more opportunity to undergo affinity maturation before transiting to the terminal plasma cell state.

At its root, the transience of the germinal center state results from the fact that IRF-4 induces Blimp-1 as well as the germinal center state. This means that, as the germinal center state is entered, a negative feedback is induced by upregulation of Blimp-1. Eventually Blimp-1 overwhelms Bcl-6 and leads to the expression of the plasma cell program at the expense of the germinal center program. Parallel dynamics form the basis for another transient differentiation event, the induction of competence for DNA uptake in *Bacillus subtilis* bacteria. This cell fate decision has been analyzed in detail both theoretically and experimentally (Suel *et al.*, 2006; Suel *et al.*, 2007). Here too, the transience is caused by the initiation of negative feedback along with the competence program (Fig. 5.7f). The key regulator of competence, comK, inhibits the expression of comS which is a positive regulator of comK necessary for its expression at levels sufficient to induce competence. Thus expression of comK drives a negative feedback loop which eventually causes the comK gene to be shut off and the cell to exit the competent state. This negative feedback loop is responsible for the transience of competence. The system differs from the B cell one in that the cycle is initiated by stochastic fluctuations in either comK or comS rather than by external stimuli (e.g., antigen binding).

One can view the specification of the germinal center and plasma cell states in the B cell model as the result of a temporal asymmetry in the expression of IRF-4. At earlier times, IRF-4 is expressed at low levels which leads to the germinal center program; at later times, IRF-4 is expressed at high levels and the cell differentiates to a plasma cell state. In other contexts, partitioning of molecules during cell division provides a mechanism for generating a spatial rather than a temporal asymmetry for cell fate specification. One example is provided by T lymphocytes following activation (Chang *et al.*, 2007). In this case, a T cell responding to antigen presented by a dendritic cell shows an asymmetric spatial partitioning of key cell fate determinants. Some cluster near the synapse that forms at the contact point between the T cell and the dendritic cell whereas others cluster at a pole on the opposite side of the cell. The mitotic spindle forms along the axis defined by these two points so that the cell fate determinants are inherited in an asymmetric fashion in the ensuing cell division. This partitioning appears to cause the daughter cells to differentiate asymmetrically: one gives rise to a memory cell and the other gives rise to an effector cytotoxic T cell. Fully accounting for these types of effects will require models that are explicitly spatial. Such spatial models have helped to shed light on the differentiation process in the *Drosophila* embryo which serves as a model system for the development of multicellular organisms (Houchmandzadeh *et al.*,

2002; Schroeder et al., 2004; Howard & ten Wolde, 2005; Janssens et al., 2006; Reeves et al., 2006; Gregor et al., 2007a; Gregor et al., 2007b).

Summary and outlook

Cell fate decisions are complex phenomena which can involve many transcription factors interacting through a complex web of interactions with feedback. Both instructive pathways acting through external signals and stochastic effects in the concentrations of the transcription factors can influence these decisions. Advances in measurement methods are now revealing specific molecular realizations of these ideas. Due to the nonintuititive nature of responses of gene regulatory networks to experimental perturbations and natural fluctuations, modeling will be essential for understanding the mechanisms governing cell fate decisions and how disruptions to the regular program of development and differentiation can give rise to pathologies.

Many binary decisions can be understood as arising from a key pair of mutually repressing transcription factors. This has been shown to be the case for both paradigmatic simple systems such as the circuit governing lysis and lyogeny in bacteria infected with phage λ (Ptashne, 1992) and complex systems such as that governing the macrophage–neutrophil decision in mice (Laslo et al., 2006). This and similar motifs are capable of responding in both graded and switch-like fashions to inputs (Biggar & Crabtree, 2001; Ozbudak et al., 2004; Laslo et al., 2006) and a cell fate decision governed by this motif can be influenced by both intrinsic and stochastic effects (Laslo et al., 2006; Huang et al., 2007). Another important regulatory motif is autoactivation which can serve as a memory of a stimulus even after it is no longer present. Such a module can be used to maintain a stable cell fate after the decision has been made by either instructive or stochastic means (Xiong & Ferrell, 2003; Acar et al., 2005). Transient differentiation events can be based on these modules and typically involve induction of a negative feedback loop simultaneously with differentiation in order to eventually repress the transient cell fate as in competence (Suel et al., 2006; Suel et al., 2007) and the germinal center program in B lymphocytes (Sciammas et al., 2006).

While a number of cell fate decisions are roughly understood and have been modeled with deterministic chemical kinetic approaches, a detailed understanding of cell fate decisions based on biologically faithful stochastic models is lacking for all but the simplest systems. This is due to both experimental and theoretical challenges which are beginning to be overcome. On the theoretical side, such models are intractable analytically and brute force numerical simulation can be prohibitively costly. In most cases, standard computational methods developed for equilibrium systems cannot be applied to these models, which involve inherently nonequilibrium processes such as transcription, protein production, and degradation. Formally exact enhanced sampling techniques for nonequilibrium systems have been tested on model systems (Allen *et al.*, 2005; Giardina *et al.*, 2006; Valeriani *et al.*, 2007; Warmflash *et al.*, 2007) and should provide a sound basis for investigating models with increasing levels of biological detail in the future. On the experimental side, there have been major advances in quantitative single-cell measurements and these techniques are just beginning to be applied to cell fate decisions in simple model organisms (Ozbudak *et al.*, 2004; Acar *et al.*, 2005; Colman-Lerner *et al.*, 2005; Suel *et al.*, 2006; Kaufmann *et al.*, 2007; Suel *et al.*, 2007). These new techniques, in combination with sophisticated molecular biological methods for perturbing complex gene regulatory networks (Medina *et al.*, 2004; Xie *et al.*, 2004; Laiosa *et al.*, 2006; Laslo *et al.*, 2006; Cobaleda *et al.*, 2007), provide insight into these processes and can be utilized for model development and validation. The combination of these theoretical and experimental developments should eventually make possible a predictive systems-level understanding of cell fate decisions in complex multicellular organisms.

Acknowledgments

We thank Harinder Singh and members of his research group for many helpful discussions. AW was supported by an NSF graduate research fellowship.

References

Abbas, A., Lichtman, A. & Pober, J. (2000). *Cellular and Molecular Immunology*. Philadelphia, PA: W. B. Saunders.

Acar, M., Becskei, A. & van Oudenaarden, A. (2005). Enhancement of cellular memory by reducing stochastic transitions. *Nature* **435**, 228–232.

Allen, R. J., Frenkel, D. & ten Wolde, P. R. (2006). Simulating rare events in equilibrium or nonequilibrium stochastic systems. *J. Chem. Phys.* **124**, 024102.

Allen, R. J., Warren, P. B. & Wolde, P. R. T. (2005). Sampling rare switching events in biochemical networks. *Phys. Rev. Lett.* **94**, 018104.

Angeli, D., Ferrell, J. E. & Sontag, E. D. (2004). Detection of multistability, bifurcations, and hysteresis in a large class of biological positive-feedback systems. *Proc. Natl. Acad. Sci. USA* **101**, 1822–1827.

Aurell, E., Brown, S., Johanson, J. & Sneppen, K. (2002). Stability puzzles in phage lambda. *Phys. Rev. E* **65**, 051914.

Bar-Even, A., Paulsson, J., Maheshri, N., *et al.* (2006). Noise in protein expression scales with natural protein abundance. *Nat. Genet.* **38**, 636–643.

Benson, M. J., Erickson, L. D., Gleeson, M. W. & Noelle, R. J. (2007). Affinity of antigen encounter and other early B-cell signals determine B-cell fate. *Curr. Opin. Immunol.* **19**, 275–280.

Biggar, S. R. & Crabtree, G. R. (2001). Cell signaling can direct either binary or graded transcriptional responses. *EMBO J.* **20**, 3167–3176.

Bintu, L., Buchler, N. E., Garcia, H. G., *et al.* (2005). Transcriptional regulation by the numbers: applications. *Curr. Opin. Gen. Dev.* **15**, 125–135.

Bolhuis, P. G., Chandler, D., Dellago, C. & Geissler, P. L. (2002). Transition path sampling: throwing ropes over rough mountain passes, in the dark. *Annu. Rev. Phys. Chem.* **53**, 291–318.

Brandman, O., Ferrell, J. E., Li, R. & Meyer, T. (2005). Interlinked fast and slow positive feedback loops drive reliable cell decisions. *Science* **310**, 496–498.

Cai, L., Friedman, N. & Xie, X. S. (2006). Stochastic protein expression in individual cells at the single molecule level. *Nature* **440**, 358–362.

Chang, J. T., Palanivel, V. R., Kinjyo, I., *et al.* (2007). Asymmetric T lymphocyte division in the initiation of adaptive immune responses. *Science* **315**, 1687–1691.

Cobaleda, C., Jochum, W. & Busslinger, M. (2007). Conversion of mature B cells into T cells by dedifferentiation to uncommitted progenitors. *Nature* **449**, 473–477.

Colman-Lerner, A., Gordon, A., Serra, E., *et al.* (2005). Regulated cell-to-cell variation in a cell-fate decision system. *Nature* **437**, 699–706.

Crouch, E. E., Li, Z., Takizawa, M., *et al.* (2007). Regulation of AID expression in the immune response. *J. Exp. Med.* **204**, 1145–1156.

Dahl, R., Iyer, S. R., Owens, K. S., *et al.* (2007). The transcriptional repressor GFI-1 antagonizes PU.1 activity through protein-protein interaction. *J. Biol. Chem.* **282**, 6473–6483.

Davidson, E. H., Rast, J. P., Oliveri, P., *et al.* (2002). A genomic regulatory network for development. *Science* **295**, 1669–1678.

Dellago, C., Bolhuis, P. G., Csajka, F. & Chandler, D. (1998). Transition path sampling and calculation of rate constants. *J. Chem. Phys.* **108**, 1964–1977.

Dias, S., Silva, H., Cumano, A. & Vieira, P. (2005). Interleukin-7 is necessary to maintain the B cell potential in common lymphoid progenitors. *J. Exp. Med.* **201**, 971–979.

Dodd, I. B., Shearwin, K. E. & Egan, J. B. (2005). Revisited gene regulation in bacteriophage lambda. *Curr. Opin. Genet. Dev.* **15**, 145–152.

Dykman, M. I., Mori, E., Ross, J. & Hunt, P. M. (1994). Large fluctuations and optimal paths in chemical kinetics. *J. Chem. Phys.* **100**, 5735–5750.

Elf, J., Li, G.-W. & Xie, X. S. (2007). Probing transcription factor dynamics at the single-molecule level in a living cell. *Science* **316**, 1191–1194.

Elowitz, M. B. & Leibler, S. (2000). A synthetic oscillatory network of transcriptional regulators. *Nature* **403**, 335–338.

Elowitz, M. B., Levine, A. J., Siggia, E. D. & Swain, P. S. (2002). Stochastic gene expression in a single cell. *Science* **297**, 1183–1186.

Erban, R., Kevrekidis, I. G., Adalsteinsson, D. & Elston, T. C. (2006). Gene regulatory networks: a coarse-grained, equation-free approach to multiscale computation. *J. Chem. Phys.* **124**, 084106.

Frenkel, D. & Smit, B. (2002). *Understanding Molecular Simulation: From Algorithms to Applications.* London: Academic Press.

Gadgil, C., Lee, C. H. & Othmer, H. G. (2005). A stochastic analysis of first-order reaction networks. *Bull. Math. Biol.* **67**, 901–946.

Gardner, T. S., Cantor, C. R. & Collins, J. J. (2000). Construction of a genetic toggle switch in *Escherichia coli. Nature* **403**, 339–342.

Giardina, C., Kurchan, J. & Peliti, L. (2006). Direct evaluation of large-deviation functions. *Phys. Rev. Lett.* **96**, 120603.

Gibson, M. A. & Bruck, J. (2000). Efficient exact stochastic simulation of chemical systems with many species and many channels. *J. Phys. Chem. A* **104**, 1876–1889.

Gillespie, D. T. (1977). Exact stochastic simulation of coupled chemical reactions. *J. Phys. Chem.* **81**, 2340–2361.

Gillespie, D. T. (2001). Approximate accelerated stochastic simulation of chemically reacting systems. *J. Chem. Phys.* **115**, 1716–1733.

Gillespie, D. T. (2007). Stochastic simulation of chemical kinetics. *Annu. Rev. Phys. Chem.* **58**, 35–55.

Glass, L. & Kauffman, S. A. (1973). The logical analysis of continuous, nonlinear biochemical control networks. *J. Theor. Biol.* **39**, 103–129.

Gregor, T., Tank, D. W., Wieschaus, E. F. & Bialek, W. (2007a). Probing the limits to positional information. *Cell* **130**, 153–164.

Gregor, T., Wieschaus, E. F., McGregor, A. P., Bialek, W. & Tank, D. W. (2007b). Stability and nuclear dynamics of the bicoid morphogen gradient. *Cell* **130**, 141–152.

Guido, N. J., Wang, X., Adalsteinsson, D., *et al.* (2006). A bottom-up approach to gene regulation. *Nature* **439**, 856–860.

Heuett, W. J. & Qian, H. (2006). Grand canonical markov model: a stochastic theory for open nonequilibrium biochemical networks. *J. Chem. Phys.* **124**, 044110.

Hock, H. & Orkin, S. H. (2006). Zinc-finger transcription factor Gfi-1: versatile regulator of lymphocytes, neutrophils and hematopoietic stem cells. *Curr. Opin. Hematol.* **13**, 1–6.

Hofer, T., Nathansen, H., Lohning, M., Radbruch, A. & Heinrich, R. (2002). GATA-3 transcriptional imprinting in Th2 lymphocytes: a mathematical model. *Proc. Natl. Acad. Sci. USA* **99**, 9364–9368.

Hornos, J. E. M., Schultz, D., Innocentini, G. C. P., *et al.* (2005). Self-regulating gene: an exact solution. *Phys. Rev. E* **72**, 051907.

Houchmandzadeh, B., Wieschaus, E. & Leibler, S. (2002). Establishment of developmental precision and proportions in the early *Drosophila* embryo. *Nature* **415**, 798–802.

Howard, M. & ten Wolde, P. R. (2005). Finding the center reliably: robust patterns of developmental gene expression. *Phys. Rev. Lett.* **95**, 208103.

Hu, M., Krause, D., Greaves, M., *et al.* (1997). Multilineage gene expression precedes commitment in the hemopoietic system. *Genes Dev.* **11**, 774–785.

Huang, S., Guo, Y.-P., May, G. & Enver, T. (2007). Bifurcation dynamics in lineage-commitment in bipotent progenitor cells. *Dev. Biol.* **305**, 695–713.

Jahnke, T. & Huisinga, W. (2007). Solving the chemical master equation for monomolecular reaction systems analytically. *J. Math. Biol.* **54**, 1–26.

Janssens, H., Hou, S., Jaeger, J., *et al.* (2006). Quantitative and predictive model of transcriptional control of the Drosophila melanogaster even skipped gene. *Nat. Genet.* **38**, 1159–1165.

Kallies, A., Hasbold, J., Tarlinton, D. M., *et al.* (2004). Plasma cell ontogeny defined by quantitative changes in Blimp-1 expression. *J. Exp. Med.* **200**, 967–977.

Kauffman, S. A. (1969). Metabolic stability and epigenesis in randomly constructed genetic nets. *J. Theor. Biol.* **22**, 437–467.

Kaufmann, B. B., Yang, Q., Mettetal, J. T. & van Oudenaarden, A. (2007). Heritable stochastic switching revealed by single-cell genealogy. *PLoS Biol.* **5**, e239.

Kikuchi, K., Lai, A. Y., Hsu, C.-L. & Kondo, M. (2005). IL-7 receptor signaling is necessary for stage transition in adult B cell development through up-regulation of EBF. *J. Exp. Med.* **201**, 1197–1203.

Klinken, S. P., Alexander, W. S. & Adams, J. M. (1988). Hemopoietic lineage switch: v-raf oncogene converts Emu-myc transgenic B cells into macrophages. *Cell* **53**, 857–867.

Kondo, M., Scherer, D. C., Miyamoto, T., *et al.* (2000). Cell-fate conversion of lymphoid-committed progenitors by instructive actions of cytokines. *Nature* **407**, 383–386.

Krysinska, H., Hoogenkamp, M., Ingram, R., *et al.* (2007). A two-step, PU.1-dependent mechanism for developmentally regulated chromatin remodeling and transcription of the c-fms gene. *Mol. Cell Biol.* **27**, 878–887.

Laiosa, C. V., Stadtfeld, M., Xie, H., de Andres-Aguayo, L. & Graf, T. (2006). Reprogramming of committed T cell progenitors to macrophages and dendritic cells by C/EBP alpha and PU.1 transcription factors. *Immunity* **25**, 731–744.

Laslo, P., Spooner, C. J., Warmflash, A., *et al.* (2006). Multilineage transcriptional priming and determination of alternate hematopoietic cell fates. *Cell* **126**, 755–766.

Lipshtat, A., Loinger, A., Balaban, N. Q. & Biham, O. (2006). Genetic toggle switch without cooperative binding. *Phys. Rev. Lett.* **96**, 188101.

Mayo, A. E., Setty, Y., Shavit, S., Zaslaver, A. & Alon, U. (2006). Plasticity of the cis-regulatory input function of a gene. *PLoS Biol.* **4**, 555–561.

Medina, K. L., Pongubala, J. M. R., Reddy, K. L., *et al.* (2004). Assembling a gene regulatory network for specification of the B cell fate. *Dev. Cell* **7**, 607–617.

Medina, K. L. & Singh, H. (2005). Gene regulatory networks orchestrating B cell fate specification, commitment, and differentiation. *Curr. Top. Microbiol. Immunol.* **290**, 1–14.

Mettetal, J. T., Muzzey, D., Pedraza, J. M., Ozbudak, E. M. & van Oudenaarden, A. (2006). Predicting stochastic gene expression dynamics in single cells. *Proc. Natl. Acad. Sci. USA* **103**, 7304–7309.

Miyamoto, T., Iwasaki, H., Reizis, B., *et al.* (2002). Myeloid or lymphoid promiscuity as a critical step in hematopoietic lineage commitment. *Dev. Cell* **3**, 137–147.

Muzzey, D. & van Oudenaarden, A. (2006). When it comes to decisions, myeloid progenitors crave positive feedback. *Cell* **126**, 650–652.

Newman, J. R. S., Ghaemmaghami, S., Ihmels, J., *et al.* (2006). Single-cell proteomic analysis of *S. cerevisiae* reveals the architecture of biological noise. *Nature* **441**, 840–846.

Novick, A. & Weiner, M. (1957). Enzyme induction as an all-or-none phenomenon. *Proc. Natl. Acad. Sci. USA* **43**, 553–566.

Orkin, S. H. (2000). Diversification of haematopoietic stem cells to specific lineages. *Nat. Rev. Genet.* **1**, 57–64.

Ozbudak, E. M., Thattai, M., Lim, H. N., Shraiman, B. I. & van Oudenaarden, A. (2004). Multistability in the lactose utilization network of *Escherichia coli*. *Nature* **427**, 737–740.

Paulsson, J. (2004). Summing up the noise in genetic networks. *Nature* **427**, 415–418.

Pedraza, J. & van Oudenaarden, A. (2005). Noise propagation in genetic networks. *Science* **307**, 1965–1969.

Ptashne, M. (1992). *A Genetic Switch*. Cambridge, MA: Blackwell Scientific Publications.

Rathinam, M., Petzold, L. R., Cao, Y. & Gillespie, D. T. (2003). Stiffness in stochastic chemically reacting systems: the implicit tau-leaping method. *J. Chem. Phys.* **119**, 12784–12794.

Reeves, G. T., Muratov, C. B., Schüpbach, T. & Shvartsman, S. Y. (2006). Quantitative models of developmental pattern formation. *Dev. Cell* **11**, 289–300.

Roma, D. M., O'Flanagan, R. A., Ruckenstein, A. E., Sengupta, A. M. & Mukhopadhyay, R. (2005). Optimal path to epigenetic switching. *Phys. Rev. E* **71**, 011902.

Rosenfield, N., Young, W., Alon, U., Swain, P. S. & Elowitz, M. B. (2005). Gene regulation at the single-cell level. *Science* **307**, 1962–1965.

Rothenberg, E. V. & Taghon, T. (2005). Molecular genetics of T cell development. *Annu. Rev. Immunol.* **23**, 601–649.

Salis, H. & Kaznessis, Y. (2005a). Accurate hybrid stochastic simulation of a system of coupled chemical or biochemical reactions. *J. Chem. Phys.* **122**, 54103.

Salis, H. & Kaznessis, Y. N. (2005b). An equation-free probabilistic steady-state approximation: dynamic application to the stochastic simulation of biochemical reaction networks. *J. Chem. Phys.* **123**, 214106.

Santillan, M. & Mackey, M. C. (2004). Why the lysogenic state of phage lambda is so stable: a mathematical modeling approach. *Biophys. J.* **86**, 75–84.

Schroeder, M. D., Pearce, M., Fak, J., *et al.* (2004). Transcriptional control in the segmentation gene network of Drosophila. *PLoS Biol.* **2**, e271.

Sciammas, R., Shaffer, A. L., Schatz, J. J., *et al.* (2006). Graded levels of interferon regulatory factor-4 coordinate isotype switching with plasma cell differentiation. *Immunity* **25**, 225–236.

Seet, C. S., Brumbaugh, R. L. & Kee, B. L. (2004). Early B cell factor promotes B lymphopoiesis with reduced interleukin 7 responsiveness in the absence of E2A. *J. Exp. Med.* **199**, 1689–1700.

Setty, Y., Mayo, A. E., Surette, M. G. & Alon, U. (2003). Detailed map of a *cis*-regulatory input function. *Proc. Natl. Acad. Sci. USA* **100**, 7702–7707.

Shalloway, D. & Faradjian, A. K. (2006). Efficient computation of the first passage time distribution of the generalized master equation by steady-state relaxation. *J. Chem. Phys.* **124**, 054112.

Sigal, A., Milo, R., Cohen, A., *et al.* (2006). Variability and memory of protein levels in human cells. *Nature* **444**, 643–646.

Singh, H., Medina, K. L. & Pongubala, J. M. R. (2005). Contingent gene regulatory networks and B cell fate specification. *Proc. Natl. Acad. Sci. USA* **102**, 4949–4953.

Singh, H., Pongubala, J. M. R. & Medina, K. L. (2007). Gene regulatory networks that orchestrate the development of B lymphocyte precursors. *Adv. Exp. Med. Biol.* **596**, 57–62.

Strogatz, S. (1994). *Nonlinear Dynamics and Chaos*. Reading, MA: Addison-Wesley Co.

Suel, G. M., Garcia-Ojalvo, J., Liberman, L. M. & Elowitz, M. B. (2006). An excitable gene regulatory circuit induces transient cellular differentiation. *Nature* **440**, 545–550.

Suel, G. M., Kulkarni, R. P., Dworkin, J., Garcia-Ojalvo, J. & Elowitz, M. B. (2007). Tunability and noise dependence in differentiation dynamics. *Science* **315**, 1716–1719.

Taghon, T., Yui, M. A. & Rothenberg, E. V. (2007). Mast cell lineage diversion of T lineage precursors by the essential T cell transcription factor GATA-3. *Nat. Immunol.* **8**, 845–855.

Takahashi, K., Kaizu, K., Hu, B. & Tomita, M. (2004). A multi-algorithm, multi-timescale method for cell simulation. *Bioinformatics* **20**, 538–546.

Tanase-Nicola, S., Warren, P. B. & ten Wolde, P. R. (2006). Signal detection, modularity, and the correlation between extrinsic and intrinsic noise in biochemical networks. *Phys. Rev. Lett.* **97**, 068102.

Thattai, M. & van Oudenaarden, A. (2001). Intrinsic noise in gene regulatory networks. *Proc. Natl. Acad. Sci. USA* **98**, 8614–8619.

Valeriani, C., Allen, R. J., Morellia, M. J., Frenkel, D. & ten Wolde, P. R. (2007). Computing stationary distributions in equilibrium and nonequilibrium systems with forward flux sampling. *J. Chem. Phys.* **127**, 114109.

Valleau, J. P. & Card, D. N. (1972). Monte Carlo estimation of the free energy by multistage sampling. *J. Chem. Phys.* **57**, 5457–5462.

van Erp, T. S., Moroni, D. & Bolhuis, P. G. (2003). A novel path sampling method for the calculation of rate constants. *J. Chem. Phys.* **118**, 7762–7774.

van Kampen, N. G. (1992). *Stochastic Processes in Physics and Chemistry*. London: Elsevier.

Vilar, J. M., Kueh, H. Y., Barkai, N. & Leibler, S. (2002). Mechanisms of noise resistance in genetic oscillators. *Proc. Natl. Acad. Sci. USA* **99**, 5988–5992.

Warmflash, A., Bhimalapuram, P. & Dinner, A. R. (2007). Umbrella sampling for nonequilibrium processes. *J. Chem. Phys.* **127**, 154112.

Warmflash, A. & Dinner, A. R. (2008). Signatures of combinational regulation in intrinsic biological noise. *Proc. Natl. Acad. Sci. USA*, in press.

Warren, P. B., Tanase-Nicola, S. & ten Wolde, P. R. (2006). Exact results for noise power spectra in linear biochemical reaction networks. *J. Chem. Phys.* **125**, 144904.

Warren, P. B. & ten Wolde, P. R. (2004). Enhancement of the stability of genetic switches by overlapping upstream regulatory domains. *Phys. Rev. Lett.* **92**, 128101.

Warren, P. B. & ten Wolde, P. R. (2005). Chemical models of genetic toggle switches. *J. Phys. Chem. B* **109**, 6812–6823.

Wylie, D. C., Hori, Y., Dinner, A. R. & Chakraborty, A. K. (2006). A hybrid deterministic-stochastic algorithm for modeling cell signaling dynamics in spatially inhomogeneous environments and under the influence of external fields. *J. Phys. Chem. B* **110**, 12 749–12 765.

Xie, H., Ye, M., Feng, R. & Graf, T. (2004). Stepwise reprogramming of B cells into macrophages. *Cell* **117**, 663–676.

Xiong, W. & Ferrell, J. E. (2003). A positive-feedback-based bi-stable 'memory module' that governs a cell fate decision. *Nature* **426**, 460–465.

Yu, J., Xiao, J., Ren, X., Lao, K. & Xie, X. S. (2006). Probing gene expression in live cells, one protein molecule at a time. *Science* **311**, 1600–1603.

6

Structural and dynamical properties of cellular and regulatory networks

R. SINATRA, J. GÓMEZ- GARDEÑES, Y. MORENO,
D. CONDORELLI, L. M. FLORÍA, AND V. LATORA

Introduction

Systems that can be mapped as networks are all around us. Recently, scientists have started to reconsider the traditional reductionism viewpoint that has driven science ever since. The accumulated evidence that systems as complex as a cell cannot be fully understood by studying only their isolated constituents, but that rather most biological characteristics and behaviors are related to complex interactions of many cellular constituents, has given rise to the birth of a new movement of interest and research in the study of complex networks, i.e., networks whose structure is irregular, complex, and dynamically evolving in time, with the main focus moving from the analysis of small networks to that of systems with thousands or millions of nodes, and with a renewed attention to the properties of networks of dynamical units. This flurry of activity has seen the physicists' and biologists' communities among the principal actors, and has been certainly induced by the increased computing powers and by the possibility to study the properties of a wealth of large databases of real networks. The regulatory and cellular networks that will be the subject of study in this chapter have been among the most studied networks, and the field has benefited from many important contributions. The expectancy is that understanding and modeling the structure of a regulatory network would lead to better comprehend its dynamical and functional behavior. In this chapter, we aim to provide the reader with a glance at the most relevant results and novel

Statistical Mechanics of Cellular Systems and Processes, ed. Muhammad H. Zaman. Published by Cambridge University Press. © Cambridge University Press 2009.

insights provided by network theory in this field, discussing both the structure and the dynamics of a number of regulatory networks.

Structure

Though in the last years many experimental techniques have been developed and larger and larger amounts of data are available, the determination and construction of the different cellular networks is not an easy task. Nevertheless, many complex interactions which take place at the cellular level, such as metabolic chains, protein–protein interactions, and gene–gene regulations, can be represented and then studied using the formalism of graph theory, where each system is reduced to a set of nodes and edges, sometimes also called vertices and links. A node usually represents a cellular constituent, like a protein or a gene. Nodes are linked by edges. An edge can be directed or undirected, whether it has a direction or not (it "goes out" of a node and "gets into" another). The meaning of edges changes case by case: in the cellular field, for instance, it could represent a reaction or an interaction between the nodes it is linking. In the next paragraphs, we describe in more details some networks of fundamental importance for cell survival. While explaining the meaning of nodes and edges in every particular context, we will also give some basic definitions of graph theory (see Figs. 6.1–6.3), in order to better understand the biological roles of graph constituents and the dynamical properties which will be discussed in the last part of this chapter.

Metabolic networks

A metabolic network is the complete set of metabolic and physical processes that determine the physiological and biochemical properties of a cell. As such, these networks comprise the chemical reactions of metabolism as well as the regulatory interactions that guide these reactions. These are accelerated, or more accurately catalyzed, by enzymes. Therefore a complete metabolic network is constituted by three kinds of nodes: metabolites, reactions, and enzymes, and by two types of edges representing mass flow and catalytic regulations. The former kind of edge links reactants to reactions and reactions to products, while the latter connects enzymes to the reactions they catalyze. Needless to say, all these edges are directed.

The above representation is not always the most suitable. In fact, it sometimes implies the assumption that some virtual intermediate complexes take place. To avoid subjective assumptions in the way the network is built up, Wagner and Fell (2001), for instance, have deduced two distinct networks from the metabolic pathways of *Escherichia coli*, which do not imply the definition of

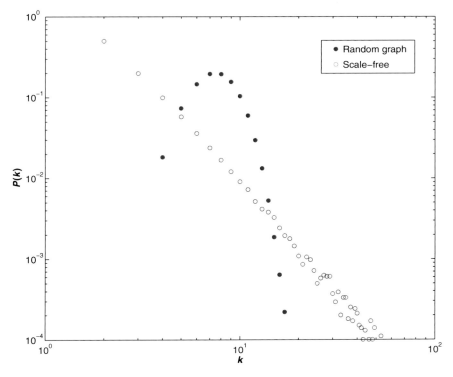

Fig. 6.1 An important graph property is the degree distribution function $P(k)$, which describes the probability to find a node with k edges. A *random graph* is constructed by randomly linking N nodes with E edges, and has a Poissonian degree distribution $P(k) = e^{-\langle k \rangle} \langle k \rangle^k / k!$. That means that the majority of nodes have a degree close to the average degree $\langle k \rangle$. A *scale-free graph* is instead characterized by a power-law degree distribution $P(k) = Ak^{-\gamma}$, usually with $2 < \gamma < 3$. A power-law distribution appears as a straight line in a double-logarithmic plot. In a scale-free graph, low degree nodes are the most frequent ones, but there are also a few highly connected nodes, usually called *hubs*, not present in a random graph.

virtual intermediate complexes. In one of the networks they define, metabolites stand for nodes which are linked to undirected edges if they participate in the same reaction. Another network is instead constituted by nodes that represent metabolic reactions linked when sharing a metabolite. The networks are respectively named metabolite network and reaction network. Both these networks exhibit a power-law degree distribution (see Figs. 6.1 and 6.2) and small-world properties (Fig. 6.3). Surprisingly, when considering the whole metabolic pathways of organisms that have evolved differently and consequently show many differences, the metabolic networks share the same topological and statistical properties, namely, those corresponding to scale-free graphs (Jeong et al., 2000).

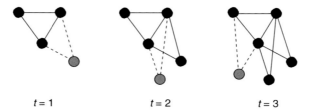

$t = 1$ $t = 2$ $t = 3$

Fig. 6.2 A simple model to grow networks with a power-law degree distribution was proposed by Barabási and Albert (1999). The model is based on two main ingredients, *growth* and *linear preferential attachment*. That means that the graph grows during time by the addition of new nodes and new links, and that links are not distributed at random, but the probability of connecting to a node depends on the nodes' degree. The algorithm to construct a network starts at time $t = 0$ with a complete graph of m_0 nodes (in the example, $m_0 = 3$). Then, at each time step t a new node n is added. The new node has $m \leq m_0$ edges (in our case, $m = 2$), linking n to m different nodes already present in the system. When choosing the nodes to which the new node n connects, it is assumed that the probability $\Pi_{n \to i}$ that n will be connected to node i is linearly proportional to the degree k_i of node i, i.e.: $\Pi_{n \to i}(k_i) = \frac{k_i}{\sum_l k_l}$. For large times (or N), this corresponds to a graph with a stationary power-law degree distribution with exponent $\gamma = 3$.

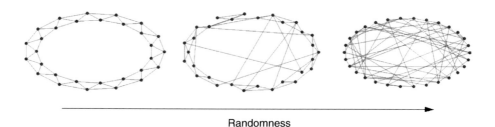

Randomness

Fig. 6.3 Small-world networks, as defined by Watts and Strogatz (1998), have intermediate properties between regular lattices (such as the first graph in the figure) and random networks (such as the last graph in the figure). A regular lattice has high clustering but also a large average path length, while a random graph is characterized by a short path length together with a low clustering. A small-world network (in the middle in the figure) borrows a high clustering coefficient from the former and a short average path length from the latter.

Protein–protein interaction networks

The interactions between proteins are crucial for many biological functions. For example, signals from the exterior of a cell are mediated to the inside of that cell by protein–protein interactions of the signaling molecules (Fig. 6.4). This process, called signal transduction, plays a fundamental role in many biological processes. Proteins may interact for a long time to form part of a protein complex, a protein may be carrying another protein, or a protein may interact briefly with another protein just to modify it. This modification of proteins can itself change protein–protein interactions. Therefore, protein–protein interactions are of central importance for every process in a living cell.

In a protein interaction network, nodes represent proteins while an undirected edge is drawn between two proteins when they physically interact. Though the data may be incomplete and contain a very high number of false positives, the results obtained from databases with very small overlap between them show the same network properties: scale-freeness, high clustering and small-world properties. These topological properties have already been exploited as the network approach allows to look at the system from new points of view and to borrow tools from other fields to solve (or at least to give alternative solutions to) known open problems. For example, this is the case of the *Saccharomyces cerevisiae* network widely studied in the literature. In particular, using the protein–protein interaction network representation of this organism (Uetz *et al.*, 2000), it has been possible to suggest or, at least, to guess the function of many unclassified proteins (Vazquez *et al.*, 2003a).

Gene regulatory networks

Some achievements in experimental techniques during the last few years, like gene chips and microarrays, have paved the way to the study of the so-called gene regulatory networks. At the cellular level, the production and degradation of all proteins are supervised by the gene regulatory network, constituted by those pairs of genes whose products in proteins are so that the former regulates the abundance of the second. That is the case, for instance, of the transcription regulatory network which is also the most studied one. It is well known that the transcription of genes from DNA to RNA is regulated by some particular proteins which are called transcription factors (see Fig. 6.4). These proteins are the products of some genes.

Therefore, the nodes of the network are genes, namely, the DNA sequences which are transcribed into the mRNAs that translate into proteins, while edges between nodes represent individual molecular reactions, the protein–protein, protein–mRNA, and protein–DNA interactions through which the products of one gene affect those of another. These interactions can be inductive, with an increase in the concentration of one leading to an increase in the other, or inhibitory, with an increase in one leading to a decrease in the other.

The method for building the graph representing a gene regulatory network is mostly based on genome-wide gene expression data. Agrawal (2002) has suggested an algorithm to build a gene regulatory network starting from microarray

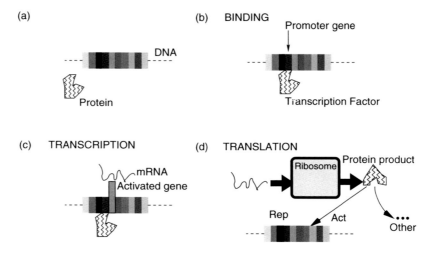

Fig. 6.4 The different stages of gene expression. (a) The basic ingredients are the proteins and specific gene regions in the DNA, such as promoters and transcribed sequences. (b) A specific protein binds to a part of the DNA sequence called the promoter; the protein is known as the transcription factor since it starts the transcription of the genetic information encoded at the specific gene that the complex promoter + transcription factor regulates. (c) After the genetic information is transcribed into the messenger RNA, by RNA polymerase, it is subsequently translated into proteins at the ribosomes (panel d). The protein product that emerges after this process can act either as another transcription factor for the expression of other genes or as a repressor of the activity of other genes stopping the synthesis of their protein products. Another possibility is that this protein product participates in the physiological processes of the cell and forms protein complexes such as enzymes. (See plate section for color version.)

gene data. For a given number N of genes, data can be represented in a matrix $N \times D$, where D is the number of sampling conditions where the expression levels appear. Looking at the rows as vectors, a Euclidean distance among genes can be defined. Each gene is then linked to its K nearest neighbors, where the Euclidean distance has been used to determine them. Many networks are then constructed by treating K as an order parameter.

Overall leitmotif

What is extremely interesting is that all the biological networks we have so far briefly described, although being very different one from the other, share many topological features. Each of them shows not only a power-law degree distribution, but also high clustering coefficient (numbers of triangles present in the network, or, in other words, the probability that if A is linked to

B and B is linked to C, A is also linked to C) and short mean path length (every couple of nodes can be connected with a path of only few links) which lead to the so-called small-worldness.

These characteristics are present not only for networks in the biological field, but also in other complex systems, going from technological to social networks. Such a universality in systems so far from one another is quite intriguing and suggests we should focus on which meanings all these topological features hide.

The power-law degree distribution involves the presence of highly connected nodes, called *hubs*, even if the small-degree ones are the most abundant (see Figs. 6.1 and 6.2). The existence of hubs seems to be correlated with evolutionary processes. The hubs should represent the oldest cellular constituents, to which new nodes, generated by gene duplication processes, preferentially attach. This growing mechanism has been shown to explain the topology of protein–protein interaction networks (Vazquez *et al.*, 2003b), but with proper adjustments it seems to explain the scale-free features of regulatory and metabolic networks as well. However, it is quite clear that highly connected nodes are subjected to severe selective and evolutionary constraints and that the cell is vulnerable to the loss of highly interactive hubs, which can result in the breakdown of the network into isolated clusters. A famous example of a hub protein is the tumor suppressor protein p53, which Vogelstein *et al.* (2000) have demonstrated to be inactive in half of human tumors.

On the other hand, the high clustering feature is related to the existence of modules (Fig. 6.5). Networks are composed by subgraphs of highly interconnected groups of nodes, usually called motifs or modules. Formally, a motif is a connected subgraph of *n* nodes which appears more frequently than in a graph with the same size, number of edges and degree distribution, but with randomized links. Each real network is characterized by its own set of distinctive motifs. It has been suggested that motifs have specific functions as elementary circuits. For example, in regulatory networks feedforward loops appear more frequently than expected from randomly connected graphs, while in protein–protein maps there is a high abundance of completely connected subgraphs and short cycles.

The molecular components constituting a motif not only interact with the elements of that motif but can be linked to other motifs giving rise to clusters and modules at a larger scale which are still interconnected to each other. Moreover, the presence of hubs makes the existence of relatively isolated modules unlikely, which ultimately gives to cellular networks a hierarchical topology. Admittedly, there is a high degree of overlap and crosstalk between modules with small modules forming cohesive communities. Interestingly, these structural modules correlate very well with functional ones, thus providing a way to study systematically the structure–function relationship.

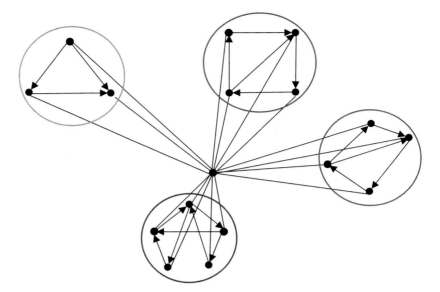

Fig. 6.5 A motif of a graph is a connected subgraph of n nodes which appears over represented if compared to a graph of the same size, number of edges and degree distribution, but with randomized connections. Motifs can group to form clusters or sometimes can interact one with each other if linked to a common hub.

Dynamics of regulatory networks

After the structural characterization of interaction maps between genes, protein and metabolites the following question turns up into scene: What is the relationship between the structure of interactions observed in most biological networks and their task-performing ability? In order to answer this question and shed new light on what is going on at the cellular and molecular levels of organization of biological systems, scientists have begun to look for the dynamical evolution of the activity patterns of the constituents of such biological networks. In fact, during the last few years, the available amount of experimental data, obtained with technological advances such as cDNA microarrays, has exploded. This has allowed the dynamical characterization of diverse biological processes both on a genome-wide and on multi-gene scales and with fine time resolution. On the other hand, despite the advances in biological engineering, the formulation of compelling models on the dynamics governing metabolic and genetic processes is still a hard issue because the observed dynamical patterns are highly nonlinear and one needs to deal with many degrees of freedom for a proper description of regulatory mechanisms.

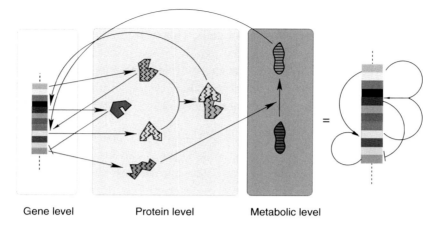

Gene level   Protein level   Metabolic level

Fig. 6.6 Coarse-graining of cellular interactions into a single gene network. The three levels of description (genes, proteins, and metabolism) and the interactions between their constituents are embedded on a single map of interactions between genes. (See plate section for color version.)

Although the ultimate goal of systems biology is to describe the cellular processes as a whole by means of a global biochemical regulatory network, the three levels of description (gene expression, protein interaction, and metabolic fluxes) are usually studied separately (as we have seen in the previous section), including when studying the dynamics of interactions. The reason behind this compartmentalization of the cellular system is the diverse ability for profiling genes, proteins, or metabolites. While current techniques for measuring genome-wide differential gene expression are nowadays widespread, this is not the case for the current methods used to deal with proteins and metabolites. In principle, one cannot get rid of the two higher organization levels of a cell; however, a number of relevant processes of cell physiology can be mapped into a gene network coarse-grained view of cells (see Fig. 6.6). In this part of the chapter we will focus on the current models used to describe the dynamics of gene regulatory networks.

As in other scientific fields, the work on the characterization of gene–gene regulation started by looking at the basic mechanisms and building blocks of the entire biological system. Following this constructionist scheme, concepts such as *operon*, *regulator gene*, and *transcriptional repression* were first introduced in the literature by Jacob and Monod (1961). Their model has settled the basis for more elaborated models as different regulatory mechanisms have been discovered (Wall *et al.*, 2004). Here, after discussing the basis of the dynamical models, we will also move from small gene circuits (so-called modules), where their predictive power is high, to large-scale gene networks, where the goal is to model the global functioning of cells. This bottom–up approach is aimed at addressing several issues of relevance in cellular processes, first on the small scale to analyze the robustness of small circuits under external (environmental) perturbations, and second on large-scale networks to test the ability of groups of genes to perform different coordinated tasks.

Mathematical formulation: Boolean models versus differential equations

Regulatory mechanisms among genes can be translated into mathematical language in various ways. The appropriate choice of the dynamical equations will depend on the level of description required. In this sense, large-scale gene regulatory networks are usually described by a simple mathematical framework that makes use of Boolean functions. On the other hand, when one is interested in describing simple regulatory mechanisms that involve few genes more detailed models such as nonlinear differential equations are best suited. The use of each type of description thus depends on the sum of complexities regarding the structure and the dynamics. In the following we describe the essential ingredients of both mathematical approaches.

Boolean modeling of regulatory networks

Boolean models are based on the assumption that genes can be found in one of a discrete set of states, and account for the different kinds of interactions that appear in gene regulatory networks by means of simple rules. Besides their simplicity, the success of Boolean models relies on the inherent difficulty in obtaining an accurate functional form of the reaction kinetics associated to every gene–gene interaction. Boolean dynamics has been widely used to analyze the importance that the global topological features of a gene network (such as path redundancy or abundance of loops, average number and sign of regulatory inputs, etc.) have on its dynamical organization.

In the usual Boolean framework, a gene i at time t can be in two possible dynamical states: *active* ($g_i(t) = 0$) or *inactive* ($g_i(t) = 1$). The activity of a gene depends on the state of those genes from which it receives a regulatory input (i.e., incoming link of the regulatory network). Besides, time is considered as a discrete variable so that at each time step the activity level of every gene i is updated considering its k_i input signals

$$g_i(t + \tau) = f_i(g_{j_1}(t), \dots, g_{j_{k_i}}(t)). \tag{6.1}$$

The updating process of the whole network can be synchronous (parallel updating) or asynchronous (sequential updating). The specific form of every function, f_i, is constructed by following the specific interactions that gene i receives from its regulators. These functions are always combinations of the basic (AND, OR,

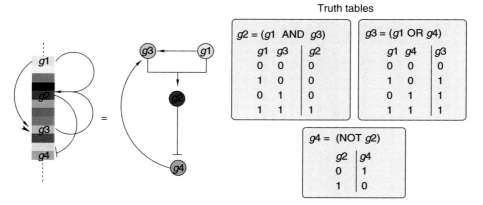

Fig. 6.7 Translation of the regulatory genetic map of Fig. 6.6 into Boolean regulatory functions. The three Boolean relations for genes g_2, g_3 and g_4 make use of the basic logical operators AND, OR, and NOT, respectively. (See plate section for color version.)

and NOT) logical operators so that the results can be either 1 if the statement is true or 0 if it is false. These single functions are expressed by means of truth tables as shown in Fig. 6.7.

Once the network and the specific Boolean functions governing the interactions are set, the study focuses on the possible dynamical behaviors. Starting from different initial conditions (in principle 2^N different possibilities) one computes how many different final states are reached. There are three possibilities, namely, (1) the system becomes frozen in a unique dynamical state (fixed point of the dynamics), (2) the system explores cyclically a set of states ending in a periodic attractor of a length given by the number of different configurations explored, (3) the dynamics is chaotic and the system explores different configurations without any periodicity. In general, different attractors can coexist in the configuration space, each of them with its own basin of attraction of initial conditions. As an example, we show in Fig. 6.8 the typical representation used to characterize the configuration dynamics of a simple regulatory network composed of three nodes, the transition between the different dynamical states reveals one periodic attractor of length 5 and the fixed point $(0, 0, 0)$.

In a large regulatory network, one expects to obtain many different types of coexisting dynamical states. This hypothesis is biologically based on the fact that different network states correspond to different cell types, i.e., cells with the same genome developing different functions. Although biochemical data from real gene networks have become available only in recent years, the search of network topologies sustaining a large number of different dynamical

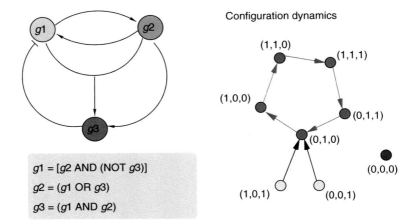

$g1 = [g2 \text{ AND } (\text{NOT } g3)]$

$g2 = (g1 \text{ OR } g3)$

$g3 = (g1 \text{ AND } g2)$

Fig. 6.8 Small regulatory network and its configuration dynamics. The network is composed of three genes that interact following the logical rules shown in the box. The configuration dynamics is represented by a network whose nodes are all the possible dynamical states and the directed links are the transitions from one to another (as dictated by the Boolean dynamics). It is shown that a cycle of length 5 (red nodes) and a steady state (blue node) exist. Yellow nodes are in the basin of attraction of the periodic attractor. (See plate section for color version.)

states started a long time ago with the pioneering work by S.A. Kauffman (1969). Kauffman considered a random assignment of the Boolean functions that governs the dynamical evolution of the gene's activity so that all the nodes receive a constant number of K input signals from other nodes and these signals are activatory or inhibitory with equal probability. The main goal was to analyze the dependence of the type, number, and length of dynamical attractors with the system size N and the number of inputs K. The results tell that for $K > 2$ the dynamics is mainly chaotic, the number of cycles scales with the number of genes, N, and their length scales exponentially with N. On the other hand, for the case $K = 1$ the dynamics is frozen and the number of attractors scales exponentially with N. Finally, the regime $K = 2$ is the most interesting since both the number and length of attractors scale as $\sqrt{N}$. These findings are very relevant biologically since the cell diversity of a living organism scales approximately with the square root of the gene number, thus pointing out that gene regulatory networks should operate just on the border between frozen and chaotic dynamics, i.e., $K_c = 2$. If one breaks the symmetry between activatory and inhibitory inputs the critical value of the network connectivity fulfills the relation

$$2\rho(1 - \rho)K_c = 1, \tag{6.2}$$

where ρ is the fraction of activatory inputs.

There has been a burst of research on Kauffman networks in the last 30 years in order to redefine models and make them more accurate (see e.g. Glass, 1975; Derrida and Pomeau, 1986; Kauffman *et al.*, 2003; Samuelson and Troein, 2003; Socolar and Kauffman, 2003; Drossel *et al.*, 2005). Perhaps, the most important refinement from the network perspective is to abandon the hypothesis of constant number of node's inputs and move to heterogeneous networks, i.e., scale-free Boolean networks. In this regard, one can consider the value K as the mean value of the number of inputs of a complex network and, in particular, re-express it as a function of the exponent γ of the power-law degree distribution. The result (Aldana and Cluzel, 2003) is that an exponent $\gamma > 2.5$ assures robust behavior (i.e., the absence of chaotic attractors) of network dynamics. Moreover, the numerical exploration (Fox and Hill, 2001) of the phase space in the regime where chaotic dynamics exist indicates that the number of observed chaotic attractors is smaller in scale-free networks than in networks with Poisson or delta degree distributions. This result could in principle relate the ubiquity of scale-free networks in regulatory systems to an evolutionary drift towards dynamical robustness.

Synthetic Boolean regulatory networks, although being idealizations, have served as test-beds for the mathematical models that are currently used on large real regulatory networks as we will see below. Besides, most of the results found for the Boolean approximation are robust when moving to more refined piecewise linear or nonlinear models. On the other hand continuous models do not allow the computation of large-scale statistical properties of their dynamical behavior.

Modeling through differential equations

Now we turn our attention to models where both time and concentrations are modeled as continuous quantities. In this case, the dynamics of the concentration of biochemical products evolves in time following the differential equation

$$\frac{d[x_i]}{dt} = f_i([x_{i_1}], [x_{i_2}], \ldots, [x_{i_n}]) - \gamma_i[x_i] , \tag{6.3}$$

where $[x]$ denotes the concentration of product x in units of *#moles/volume*. The second term in the right-hand side of (6.3) accounts for the degradation of x_i, being γ_i the degradation rate parameter. The functional form of f_i, i.e., the rate of the reaction that produces x_i, is dictated by the reaction kinetics of the chemical processes at work. This makes it extremely difficult to obtain a general framework since interactions of a gene, its products, and its regulators are reaction-specific; in fact a whole expression should contain various biochemical processes such as reaction reversibility, product dimerization, enzyme catalysis, etc.

The most simple examples of gene regulation are those of a gene x regulated by a transcription factor y acting either as a repressor or as an activator. In these cases it is easy to show (Alon, 2007) that the input functions to be included in (6.3), are respectively for repression and activation:

$$f_x([y]) = \frac{\beta}{1 + K^{-n}[y]^n} \quad \text{(Repression)}, \tag{6.4}$$

$$f_x([y]) = \frac{\beta[y]^n}{1 + K^{-n}[y]^n} \quad \text{(Activation)}. \tag{6.5}$$

In both expressions, $[x]$ is the concentration of mRNA transcribed by gene x, and β is the maximal rate of mRNA transcription so that these expressions can be read as β times the probability that the gene promoter DNA region is either free (occupied) by the repressor (activator) transcription factor y. The parameter K is the dissociation constant of the reaction that describes the binding of y to the gene promoter in DNA. The value of n (the so-called Hill coefficient) accounts for the number of transcription factor subunits that binds to the promoter. The repression input function shows that the gene activity decreases to zero as the repressor concentration grows. On the other hand the activatory input is a growing and saturable function, and when $n = 1$, it takes the expression of the Michaelis–Menten equation ubiquitously found for a wide range of biological processes such as enzyme kinetics (Sethna, 2006).

These mechanisms of negative or positive regulation of a gene can be closed in a negative or positive feedback loop if we consider that the transcription factor that regulates gene activity is the same as the protein product of mRNA translation. In this case a second differential equation

$$\frac{d[y]}{dt} = \alpha[x] - \gamma_y[y] \tag{6.6}$$

accounts for the rate production of protein y regulated by the activity of gene x. This equation has to be solved together with (6.3) with the input function given in (6.4) or (6.5). The solution to the feedback loop depends on whether there is activation or inhibition of y to x as is shown in Fig. 6.9. For the negative regulation a unique steady state exists and it is linearly stable meaning that any small perturbation of the system will return to the original state (homeostasis). On the other hand, when positive regulation occurs three possible steady states are possible, with only two of them being stable. In this case the system can choose between two different cell states.

The above modeling for gene activity level can be extended to the more realistic case when there is more than one transcription factor regulating the gene's expression. What are the effects of a activators and r repressors coordinated in the gene dynamics? In the simple case where different proteins can bind all together in the promoter region and protein complexes are not formed after

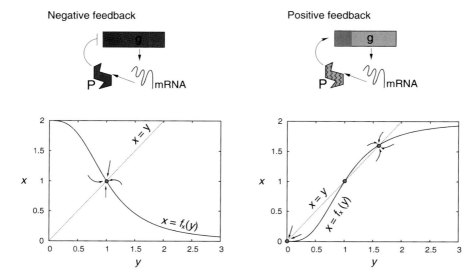

Fig. 6.9 Negative (left) and positive (right) feedback loops. Both systems describe a gene encoding its own transcription (repressor or activator) factor. Solving graphically the two systems of coupled differential equations, equations (6.4)–(6.6) and (6.5)–(6.6), one can compute the steady states of the system. In the first case, dynamics of mRNA and protein concentrations reaches a unique stable fixed point. For the positive loop there are three steady states, two of them being stable (and thus biologically reliable), (one corresponding to the rest state of the system). We have set in both cases $\beta = 2$, $n = 3$ and $\gamma_x = \gamma_y = \alpha = K = 1$.

translation, one must consider all the possible configurations for the complex *(gene promoter + transcription factors)* that allow gene expression and sum up their associated probabilities. A first-order approximation to the general formula is given by this general input gene function (Alon, 2007):

$$f_x([y_1], \ldots, [y_a], [y_{a+1}], \ldots, [y_{a+r}]) = \frac{\sum_{i=1}^{a} \beta_i([y_i]/K_i)^{n_i}}{1 + \sum_{i=1}^{a+r} \beta_i([y_i]/K_i)^{n_i}}, \tag{6.7}$$

where we have ordered the arguments of $f_x(\mathbf{y})$ so that the first a variables correspond to activators and the remaining r are repressors. A more detailed construction (where, e.g., protein dimerization is taken into account) of a mean-field model for gene regulatory inputs can be found in Andrecut and Kauffman (2006).

Although the above mathematical setting is only suited when small network circuits or modules (see below) are analyzed, one can relax the rigidity of regulatory input functions and construct more general equations that incorporate the main features of regulatory dynamics as the saturable character of the gene response under activatory inputs. In Gómez-Gardeñes *et al.* (2006) the authors make a coarse-grained formulation of continuous time gene dynamics by writing the following equations for mRNA concentrations,

$$\frac{d[x_i]}{dt} = -[x_i] + \beta \frac{\Phi\left(\sum_{i=1}^{N} W_{ij}[x_j]\right)}{1 + K^{-1}\Phi\left(\sum_{i=1}^{N} W_{ij}[x_j]\right)}, \tag{6.8}$$

where $\mathbf{W}$ is the interaction matrix whose entries are $W_{ij} = 1$ if product of gene j activates expression of gene i, $W_{ij} = -1$ if product of gene j inhibits gene i, and $W_{ij} = 0$ if no regulatory interaction is found. Besides $\Phi(x)$ is defined as $\Phi(x) = x$ if $x > 0$ and $\Phi(x) = 0$ otherwise. This compact and simple form for regulatory continuous dynamics paves the way for a thorough study of two essential ingredients in biological regulatory networks: saturability of the interactions and scale-free character of the interconnections among constituents.

Inspired by the aforementioned works in Kauffman Boolean networks synthetic Barabási–Albert scale-free networks are constructed (see Fig. 6.2) and the sign of every interaction is assigned so that the fraction of inhibitory inputs is equal to $p \in [0, 1]$. The (continuous) phase space of the systems is then explored by analyzing the final dynamical attractor of many different initial conditions. As in the studies of Boolean synthetic networks, the results here point out that different regimes (steady states, periodic, and chaotic attractors) are possible depending on the value of p (which plays here the role of ρ in Boolean networks) as seen in Fig. 6.10.

In addition, there are other dynamical features intrinsically related to the continuous character of the equations. In particular, large networks can dynamically fragment so that topologically disconnected subsets of nodes sustain independent dynamics (e.g., steady states of nonzero activity and periodic dynamics) while the rest of the network's nodes remain in the rest state. The observation of such clusters of active nodes provides an emergent network topology that is defined by those connected nodes sharing the same kind of dynamics and the links between them. The topological analysis of these dynamical clusters reveals remarkable differences from the substrate scale-free network topology such as the existence of a high clustering coefficient. This latter result is much in agreement with the topological analysis of real biological networks, and points out that the observed topology is a result of functional (dynamical) relations between elements and therefore interpretations about the origin of network patterns should not disregard dynamical analysis. In the next section we will see how small network subunits (analogous to these dynamical clusters)

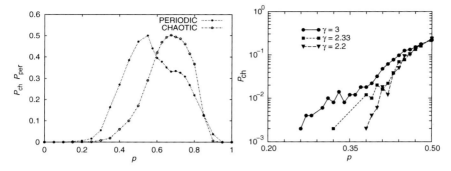

Fig. 6.10 (Left panel) Probabilities that an arbitrary initial condition ends in a periodic, P_{per}, and chaotic attractor, P_{ch}, as functions of the fraction of inhibitory inputs in the network, p. Obviously when $p = 1$ only zero-activity states are achieved thus both probabilities tend to zero as $p \to 1$. The results show that there is a region where dynamical order (periodic states or steady states) prevails on the network dynamics, and that the threshold of order (no chaotic states) is around $p_c \simeq 0.26$. This threshold (right panel) seems to grow slightly when the exponent γ of the power-law degree distribution decreases (and so the degree of heterogeneity in the network grows). After Gómez-Gardeñes *et al.* (2006).

with robust and coherent dynamics, termed motifs, have attracted a lot of attention when studying real regulatory networks.

Dynamics of real regulatory networks

We now focus on applications to modeling real gene regulatory networks. As discussed in the previous section, an important issue concerns the most convenient level of description for a particular network (Bornholdt, 2005). While network subunits or modules can be modeled in terms of differential equations, a description of an entire regulatory system needs to rely on the coarse-grained picture of logical or Boolean dynamics.

Network modules

Network modules are presented in the literature as small circuits (composed typically of three genes) embedded in large regulatory networks that are able to display autonomous dynamics. Since it is difficult to detect modules by simply looking at the whole network activity, several works have focused on the identification of general building blocks in gene networks by looking for motifs in the network topology (Milo *et al.*, 2002). Motifs (as previously noted) are those subgraphs whose occurrence in the real network is significantly higher than in their randomized versions and include autoregulatory excitatory feedback loops, inhibitory feedback loops, feedforward loops, and dual positive-feedback

loops. Once network motifs are identified, different tests of dynamical robustness or reliability (e.g., synchronizability) in different regulatory networks can
be performed (Bradman *et al.*, 2005; Klemm and Bornholdt, 2005; Ma'ayan *et al.*,
2005; Lodato *et al.*, 2007). When regulatory dynamics is implemented, the result
is that network motifs show a more robust behavior than other circuits, with
the same number of genes, that are not so frequent in the regulatory map. As
a consequence, the experimental occurrence of particular structures of regulatory interactions seems to be due to their remarkable dynamical reliability.
This suggest a selective process acting on the *pattern of interactions* rather than
on *isolated* genes.

The finding that a few basic modules are the building blocks of large real
regulatory networks justifies the design and construction of small synthetic
regulatory circuits to implement particular tasks. The most salient example of
a synthetic gene network is the *repressilator* that has become one of the best-
studied model systems of this kind. The repressilator is a network of three
genes, whose products (proteins) act as repressors of the transcription of each
other in a cyclic way (see Fig 6.11). This synthetic network was implemented
in the bacterium *E. coli* so that periodically it induces the synthesis of a green
fluorescent protein as a readout of the repressilator state (Elowitz and Leibler,
2000). In this regard, the temporal fluctuations in the concentration of each of
the three components of the repressilator can be easily reproduced by analyzing

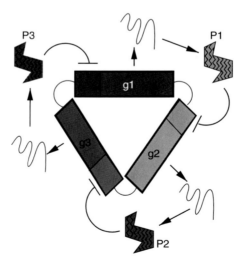

Fig. 6.11 Schematic representation of the repressilator. The repressilator is a small
network composed of three genes g_i ($i = 1, 2, 3$), each one inhibiting the activity of
the subsequent. That is, the protein products of genes g_1, g_2, and g_3 act as the
repressors of the activity of genes g_2, g_3, and g_1 respectively.

a system of six ordinary differential equations, based on (6.3, 6.4, and 6.6), reading

$$\frac{d[x_i]}{dt} = -[x_i] + \frac{\beta}{1 + [y_j]^n} , \qquad \text{(mRNA dynamics)} \tag{6.9}$$

$$\frac{d[y_i]}{dt} = -\alpha([y_i] - [x_i]) , \qquad \text{(protein dynamics)} \tag{6.10}$$

where the couples (i, j) assume the values $(1, 3)$, $(2, 1)$, and $(3, 2)$. The variable $[x_i]$ is the mRNA concentration encoded by gene x_i, and $[y_i]$ is the concentration of its translated protein y_i. The parameter α is the ratio of the protein decay rate to the mRNA decay rate, and time has been rescaled in units of the mRNA lifetime. This system of equations has a unique steady state which can be stable or unstable depending on the parameter values. In the unstable region of parameter space, the three protein concentrations fluctuate periodically. Experiments show the temporal oscillations of fluorescence, which were checked to be due to the repressilator, validating the model predictions. In particular, the previous mathematical model served to identify possible classes of dynamical behavior and to determine what experimental parameters should be adjusted in order to obtain sustained oscillations.

The repressilator is an illustrative example of the experience gained by identifying network modules and modeling its dynamical behavior in real networks. Not surprisingly, the repressilator called attention from experts on (biological) synchronization, for it offers good prospects for further insights into the nature of biological rhythms, whose mechanisms remain to be understood. In this respect, a simple modular addition of two proteins to the repressilator original design has been recently proposed (Garcia-Ojalvo et al., 2004). This extension is made so that one of the new proteins can diffuse through the cell membrane thus providing a coupling mechanism between cells containing repressilator networks. This intercell communication couples the dynamics of the different cell oscillators (with different repressilator periods) and thus allows the study of the transition to synchronization of coupled phase oscillators in a biological system. The result reproduces the phase transition from uncorrelated to coherent dynamics as the cell dilution decreases (increasing cell–cell interaction).

Large regulatory networks

The study of elementary gene circuits certainly provides answers to intriguing questions about the regulatory mechanisms at work and their organizational patterns. On the other hand, the description of how the organization at the system level emerges is far from being trivial. Despite the enormous complexity differences between such living organisms as worms (e.g., *Caenorhabditis elegans*) and humans the difference between their genome length is too small to explain the species' evolutionary gap. The difference should be unveiled by looking at the complexity of each gene network that relies on the variety of collective responses or phenotypes it displays. It is therefore clear that genome-wide approaches will allow to discover new higher-order patterns.

The characterization of the dynamical complexity of real regulatory networks is a very recent issue. The most successful approach is to use Boolean dynamics to characterize the regulatory interactions and thus to construct an oversimplified (free of parameters) model. Recent studies on this direction have addressed different regulatory networks, some examples are found in references (Mendoza *et al.*, 1999; Albert and Oltmer, 2003; Li *et al.*, 2004; Faure *et al.*, 2006; Davidich and Bornholdt, 2007). Although the number of gene networks that are currently analyzed from the dynamical point of view is growing, it will still take a long time to follow the evolution of complexity in living organisms by comparing gene networks and their dynamical behavior. Up to now the models focus on reproducing the sequential expression patterns observed experimentally. This is the case for the segment polarity gene network in *Drosophila melanogaster* (Albert and Oltmer, 2003), and the cell cycles of *S. cerevisiae* (Li *et al.*, 2004) and *Schizosaccharomyces pombe* (Davidich and Bornholdt, 2007). These two latter studies are particularly important since the two cells are well-studied eukaryotic organisms and their network dynamics show remarkable differences. In particular, it has been observed (Davidich and Bornholdt, 2007) that although both cycles are similar in terms of the size of the basin of attraction, the overall dynamics observed when external signals are introduced is qualitatively different.

More promising, from the point of view of predicting power, are the findings for the flower morphogenesis regulatory network in *Arabidopsis thaliana* [28]. (Mendoza *et al.*, 1999). In this case the five different phenotypes (dynamical attractors) of the flower (petals, sepals, stamen, carpels, and flower inhibition) are reproduced, plus a new sixth phenotype that does not correspond to any previously found cell type. The above examples summarize the large predicting power of dynamical models in large real regulatory networks. Despite the use of a coarse-grained view such as Boolean dynamics, the qualitative aspects of cellular dynamics seem to be already captured in this framework.

Conclusions

Finally, we would like to emphasize that the studies mentioned here and others available in the literature are only the tip of the iceberg. It is expected that new tools will come into play and that the universal behavior observed in the topology of many diverse phenomena from physical, social, technological and biological systems will allow a cross-fertilization between different disciplines, the ultimate goal being to tackle the complex structural and dynamical relationships in living systems such as the cell. If this is achieved, then we will be able to use that consistent framework to make predictions and to develop alternative experimental techniques and practical applications such as targeted drugs.

References

Agrawal, H. (2002). Extreme self-organization in networks constructed from gene expression data, *Phys. Rev. Lett.* **89**, 268702.

Albert, R. (2005). Scale-free networks in cell biology, *J. Cell Sci.* **118**, 4947–4957.

Albert, R. and Oltmer, H. G. (2003). The topology of the regulatory interactions predicts the expression pattern of the *Drosophila* segment polarity genes, *J. Theor. Biol.* **223**, 1–18.

Aldana, M. and Cluzel, P. (2003). A natural class of robust networks, *Proc. Natl. Acad. Sci. USA* **100**, 8710–8714.

Alon, U. (2007). *An Introduction to Systems Biology: Design Principles of Biological Circuits.* Chapman & Hall.

Andrecut, M. and Kauffman, S. A. (2006). Mean-field model of genetic regulatory networks, *New J. Phys.* **8**, 148.

Barabási, A. L. and Albert, R. (1999). Emergence of scaling in random networks, *Science* **286**, 509–512.

Barabási, A. L. and Oltvai, Z. (2004). Network biology: understanding the cell's functional organization, *Nature Rev. Genetics* **5**, 101–113.

Bornholdt, S. (2005). Less is more in modeling genetic networks, *Science* **310**, 449–451.

Bradman, O., Ferrel Jr., J. E., Li, R., and Meyer, T. (2005). Interlinked fast and slow positive feedback loops drive reliable cell decisions, *Science* **310**, 496.

Davidich, M. I., and Bornholdt, S. (2007). Boolean network model predicts cell-cycle sequence of fission yeast, *arXiv/q-bio*, 0704.2200.

Derrida, B. and Pomeau, Y. (1986) Random networks of automata: a simple annealed approximation, *Europhys. Lett.* **1**, 45–49.

Drossel, B., Mihaljev, T., and Greil, F. (2005). Number and length of attractors in a critical Kauffman model with connectivity one, *Phys. Rev. Lett.* **94**, 88 701.

Elowitz, M. B., and Leibler, S. (2000). A synthetic oscillatory network of transcriptional regulators, *Nature* **403**, 335–338.

Faure, A., Naldi, A., Chaouiya, C., and Thieffry, D. (2006). Dynamical analysis of a generic Boolean model for the control of the mammalian cell cycle, *Bioinformatics* **22**, e124–e131.

Fox, J. J. and Hill, C. C. (2001). From topology to dynamics in biochemical networks, *Chaos* **11**, 809–815.

García-Ojalvo, J., Elowitz, M. B., and Strogatz, S. H. (2004). Modeling a synthetic multicellular clock: repressilators coupled by quorum sensing, *Proc. Natl. Acad. Sci. USA* **101**, 10 955–10 960.

Glass, L. (1975). Classification of biological networks by their qualitative dynamics *J. Theor. Biol.* **54**, 85–107.

Gómez-Gardeñes, J., Moreno, Y., and Floría, L. M. (2006). Scale-free topologies and activatory-inhibitory, interactions, *Chaos* **16**, 15114.

Jacob, F. and Monod, J. (1961). Genetic regulatory mechanisms in the synthesis of proteins, *J. Mol. Biol.* **3**, 318–356.

Jeong, H., Tombor, B., Albert, R., Oltvai, Z. N., and Barabasi, A. L. (2000). The large-scale organization of metabolic networks, *Nature* **407**, 651.

Kauffman, S. A. (1969). Metabolic stability and epigenesis in randomly constructed genetic sets, *J. Theor. Biol.* **22**, 437–467.

Kauffman, S. A., Peterson, C., Samuelsson, B., and Troein, C. (2003). Random Boolean network models and the yeast transcriptional network *Proc. Natl. Acad. Sci. USA* **100**, 14 796–14 799.

Klemm, K. and Bornholdt, S. (2005). Topology of biological networks and reliability of information processing, *Proc. Natl. Acad. Sci. USA* **102**, 18 414–18 419.

Li, F., Long, T., Lu, Y., Quyang, Q., and Tang, C. (2004). The yeast cell-cycle network is robustly designed, *Proc. Natl. Acad. Sci. USA* **101**, 4781–4786.

Lodato, I., Boccaletti, S., and Latora, V. (2007). Synchronization properties of network motifs, *Europhys. Lett.* **78**, 28 001.

Ma'ayan, A. *et al.* (2005). Formation of regulatory patterns during signal propagation in a mammalian cellular network, *Science* **309**, 1078–1083.

Mendoza, L., Thieffry, D., and Alvarez-Buylla, E. R. (1999). Genetic control of flower morphogenesis in *Arabidopsis thaliana*: a logical analysis, *Bioinformatics* **15**, 593–606.

Milo, R., Shen-Orr, S., Itzkovitz, S., Kashtan, N., Chklovskii, D., and Alon, U. (2002). Network motifs: simple building blocks of complex networks, *Science* **298**, 824–827.

Samuelsson, B., and Troein, C. (2003). Superpolynomial growth in the number of attractors in Kauffman networks, *Phys. Rev. Lett.* **90**, 98 701.

Sethna, J. P. (2006). *Statistical Mechanics: Entropy, Order Parameters, and Complexity.* Oxford University Press.

Socolar, J. E. S. and Kauffman, S. A. (2003). Scaling in ordered and critical random Boolean networks, *Phys. Rev. Lett.* **90**, 68702.

Uetz, P. *et al.* (2003). A comprehensive analysis of protein–protein interactions in *Saccharomyces cerevisiae*, *Nature* **403**, 623.

Vazquez, A., Flammini, A., Maritan, A., and Vespignani, A. (2003a). Global protein function prediction from protein–protein interaction networks, *Nature Biotech.* **21**, 697.

Vazquez, A., Flammini, A., Maritan, A., and Vespignani, A., (2003b). Modeling of protein interaction networks, ComplexUs 1, 38.

Vogelstein, B., Lane, D., and Levine, A. J. (2000). Surfing the p53 network, *Nature* **408**, 307–310.

Wagner, A. and Fell, D. A. (2001). The small world inside large metabolic networks, *Proc. R. Soc. London Ser. B* **268**, 1803–1810.

Wall, M. E., Hiavacek, W. S., and Savageau, M. A. (2004). Design of gene circuits: lessons from bacteria, *Nature Rev. Genetics* **5**, 34–42.

Watts, D. J. and Strogatz, S. H. (1998). Collective dynamics of "small-world" networks, *Nature* **393**, 440–442.

7

Statistical mechanics of the immune response to vaccines

JUN SUN AND MICHAEL W. DEEM

7.1   Introduction

The immune system provides a biological example of a random, disordered, and evolving system. A faithful model of the immune system must capture several unique biological features. First, the antibodies of the immune system each contain many amino acids with many interactions among them, yet the amino acids are of 20 types. Second, each person has only a finite number of antibodies, out of a nearly infinite space of possible antibodies. Third, dynamics is important, because diseases must be cleared from an infected person via the immune system relatively rapidly. Fourth, there are correlations between the immune system and the pathogen, because diseases evolve in response to immune system pressure. Due to these four technical features, it is natural to use statistical mechanics to build quantitative models of the immune system, in an effort to understand the response of the immune system to pathogens and to optimize this response via vaccine design.

Although a significant number of detailed experiments have been performed on the immune system, it is far from completely understood. New scientific discoveries can have significant impacts on public health, as many diseases remain banes to mankind. For example, the worldwide mortality from influenza epidemics is 250 000 to 500 000 people annually, with 5% to 15% of the world's population becoming ill each year [1]. Typical costs associated with influenza in the USA are $10 billion annually [2], with approximately 40 000 deaths, and

Statistical Mechanics of Cellular Systems and Processes, ed. Muhammad H. Zaman. Published by Cambridge University Press. © Cambridge University Press 2009.

this figure increases to approximately 90 000 if individuals with complicating medical conditions such as heart disease are included [3, 4]. The economic cost of a possible influenza pandemic in the USA has been estimated by the US Centers for Disease Control to be between $71 and $167 billion [5]. It is stated by the World Health Organization that vaccination is the primary public health method used to prevent infection with influenza [6]. Another viral disease that poses a challenge to global health is dengue, and dengue virus infections are a significant cause of morbidity and mortality in tropical and subtropical regions [7–9]. There are an estimated 100 million cases of dengue annually [7]. A serious secondary form of the disease, dengue hemorrhagic fever, leads to the hospitalization of several hundred thousand people each year, almost always after a previous infection with a different serotype of dengue [7, 10, 11].

Six of the 14 "Grand Challenges in Global Health" are vaccine-related, and quantitative models of the immune system address two of these Grand Challenges [12]. The average efficacy of the influenza vaccine over the past 35 years has been about 41% of the typical value for a perfectly designed vaccine, both due to the mutation of the flu and to the less than optimal design of the vaccine [13]. The efficacy of the flu vaccine has even been negative in about 26% of the years [13]. Two of the most well-known diseases with high death rates, cancer and HIV, are immunity related, and both raise challenges to vaccine design due to their high mutation rates and diversities. Autoimmunity, that is the situation in which an immune response is mistakenly directed toward one's own proteins, is a poorly understood malfunction of immunology that is affecting an increasing number of people. Autoimmune diseases currently affect some 50 million Americans, are one of the top ten leading causes of death in women aged 65 years and younger and in children, and cost approximately $100 billion in direct healthcare costs annually in the USA [14]. The development of quantitative models of the immune system response to vaccines, viruses, and other antigens would seem to be a productive approach to study of the immune system and useful contribution to public health.

The antibodies of the immune system evolve during an immune response. This evolution occurs in the sequence space of the antibodies that recognize the disease. The binding energy landscape of the antibody to the antigen is rugged due to randomness, disorder, and many-body effects, so the landscape depends on the amino acid sequence in an unpredictable way. The mechanism by which sequences evolve is point mutation or local diffusion in the space of possible sequences. We therefore expect evolution of antibodies to be generically slow and perhaps glassy, due to the random landscape upon which the antibodies evolve and are selected for. Thus, we are motivated to model the protein evolution that occurs in the immune response with mutation dynamics on a spin glass model [15]. Since protein folding occurs on a significantly more rapid timescale than does the evolutionary dynamics that we consider, we here focus on the evolution of the amino acid sequence degrees of freedom.

Our work complements the traditional literature of mean-field, differential equation based approaches to the immune system in two aspects. First, we investigate the immune response of a finite size immune system. The finite diversity of the finite size antibody repertoire, therefore, is critical. Diversity reduction in the antibody repertoire may lead to original antigenic sin. The immune system dynamics of variable diversity joining (VDJ) recombination and somatic hypermutation are captured by the mutation and selection dynamics of our model. Exactly these glassy evolutionary dynamics are what our model can capture, and what differential equation based models cannot. Autoimmunity may occur due to the glassy nature of the immune system and chronic pathogen exposure. The sequence-based approach we describe captures the finite diversity and the glassy dynamics of the immune system, which are not intrinsically included in mean-field theories.

That proteins are composed from smaller domains and secondary structures induces a hierarchical structure to the ruggedness of the antibody-binding energy in the amino acid sequence space in which antibodies evolve. These correlations imply that while the dynamics of antibody evolution may be glassy, the speed of evolution depends on whether the evolutionary dynamics is complementary to the hierarchical structure. To describe the biology properly, it is essential to take into account this hierarchical nature of the evolution. This hierarchical structure is a novel feature of the spin glass model that we discuss. In nature there exists a wide range of evolutionary mechanisms [16–19], which range from point mutation to recombination to genetic swapping events, and this hierarchy is exploited in evolution in the immune system as well [20]. In the immune system, VDJ recombination has evolved to provide diversified initial starting sequences for antibodies, and somatic hypermutation has evolved to fine tune these sequences.

The exposure of naive B or T cells of the immune system to a pathogen may lead to memory. The memory sequences help to start a quick and effective response when the second antigen is similar, but it reduces the diversity of the antibody repertoire. We uniquely predict with our spin glass model the phenomenon of original antigenic sin, i.e., under some conditions vaccination creates memory sequences that increase susceptibility to future exposure to the same disease. Our prediction of vaccine efficacy versus antigenic distance based on dominant epitope fits the epidemiological data very well. We tackle autoimmunity with our model. Cross-reaction occurs when antibody raised against antigen A also binds antigen B. This is one of the mechanisms by which autoimmune disease can develop. We suggest that to avoid autoimmunity a balance has evolved between affinity and specificity in searching sequence space. Our T cell model captures the realistic recognition characteristics between T cell receptors and virus at the sequence level, and predicts dengue vaccine clinical trial data well. We propose polytopic vaccination as a means to overcome the challenges of immunodominance caused by the four closely related serotypes of dengue fever.

The rest of this chapter is organized as follows. In Section 7.2, we provide a brief description of the immune system. In Section 7.3, we describe the sequence-based, hierarchical spin glass model that we use to describe antibody evolution in the immune system. We also present the model of T cell evolution. In Section 7.4, we choose several examples to apply this model to immunological problems: we describe the phenomenon and mechanism of original antigenic sin; we describe a new order parameter derived from the theory that correlates well with efficacy of the annual flu shot; we describe how the glassy nature of evolution within the immune system helps to avoid autoimmune disease; and we describe how the T cell model is used to discuss treatment of dengue fever by polytopic vaccination. We conclude in Section 7.5. Detailed versions of some of these results have appeared previously [13, 21–24].

7.2   Brief description of the immune system

The immune system protects us from infection due to invading pathogens. The immune system contains cells and molecules circulating in the fluid of our tissues that interact through signaling to locate and eliminate invaders. There are two subsystems of the immune system, the innate and the adaptive immune system. The innate immune system recognizes invading foreign bodies using a general mechanism, whereas the adaptive immune system recognizes pathogens in a specific way. There are two further categories of the adaptive immune response, namely the B cell response and the T cell response. The B cell, also called humoral, immune response is dedicated to the elimination of extracellular pathogens, while the T cell, also called cellular, response focuses on intracellular pathogens.

The main action of the humoral immune response is to create and evolve B cells that produce antibodies specific for antigen. An antibody molecule consists of two identical light and two identical heavy protein chains. B cells that produce an antibody which binds strongly to an antigen begin to divide. This division occurs in a lymph node, of which there are many in the body, with a rate of several divisions per day.

Antibodies are produced with wide range of diversity and in massive number. The diversity of antibodies results from the process called VDJ recombination, in which genetic fragments coding for pieces of antibodies are randomly selected from three different regions in a person's genome (V, D, and J) are joined together. This process creates an initial set of antibodies. On the order of 10^{12} to 10^{14} different antibodies can be produced in this way from one genome, although only 10^8 are present in any one person at any one point in time [20]. B cells producing antibodies recognizing an invading pathogen undergo the second evolutionary process of somatic hypermutation. In somatic hypermutation, B cells producing antibodies binding to the antigen with high affinities are selected for and expanded in number by division. By this process, the DNA coding for the antibody-binding region is also mutated. Somatic hypermutation performs a small-scale search in the amino acid sequence space, through the action of individual point mutations, for antibodies of higher affinity, with the starting point of the search the initial VDJ recombinants [25]. One can view somatic hypermutation as local diffusion in the space of the antibody sequences.

The binding of antibody to antigen is a reversible chemical reaction:

$$K^{eq} = \frac{[\text{Antibody} : \text{Antigen}]}{[\text{Antibody}][\text{Antigen}]}, \tag{7.1}$$

where $[\cdot]$ indicates concentration. In nature the binding constant K ranges from 10^4 to 10^7 l/mol, whereas in experimental antibody engineering studies the binding constant can exceed 10^{13} l/mol [20, 26].

Some of the B cells that produce antibodies that recognize antigen differentiate into plasma cells and others become memory cells. Plasma cells secrete effector antibodies soluble in the fluid of the body's tissues. These antibodies bind antigen and block them from entering cells, signal phagocytic cells to ingest and destroy the invaders, or activate the complement to punch holes in the membranes of the pathogen. Memory cells tend to stay at peripheral areas, such as lymph nodes, and their membrane-bound antibodies have the same specificity as the effector antibodies. Memory cells can help the immune system to mount a rapid and effective response when the host encounters a pathogen similar to the pathogen (or vaccine) that stimulated the production of the memory B cells. The memory mechanism is, therefore, the foundation of vaccination.

The T cell immune response finds T cell receptors (TCRs) that bind specifically to the antigenic peptides presented by major histocompatibility complex (MHC) molecules. TCRs only recognize peptide presented by MHC molecules, and the TCR, MHC, and antigenic peptide form a three-body complex. MHC molecules only bind to short peptides, i.e., hydrolyzed antigen, of eight to ten amino acids. There are two kinds of TCR, CD4 and CD8, binding to MHCII and

MHCI molecules. We will focus on the dynamics of CD8+ TCRs in our simplified discussion of the immune system.

T cells originate in bone marrow and mature in the thymus. They acquire their diversity through VDJ recombination. During the development of TCRs, they undergo rounds of selection for increased avidity. They do not go through somatic hypermutation, as do antibodies, presumably because further evolution may yield TCRs with unnecessarily high affinity and cross-reactivity against other short nine amino acid peptides present in the body and cause autoimmune disease. Some mature T cells proliferate and produce effector T cells, whereas others become memory cells.

7.3   Hierarchical spin glass model of antibody or T cell receptor evolution

B cell model

We use the generalized *NK* model to study the interactions between antibodies and antigen [21, 27]. This model is a hierarchical spin glass model. The model captures both the correlations and the ruggedness of the interaction energy between the antibody and the antigen. The space of variables is that of the antibody amino acid sequence and the composition of the antigen. The correlations are due to the physical structure of the antibodies and the antigen. The hierarchical structure of this model distinguishes it from traditional long- or short-ranged spin glass models [28, 29]. We are studying the immune and antigen system at the sequence level. There are two reasons for this decision. One reason is that study at the atomic level is currently an intractable problem, with 10^4 atoms per antibody, 10^8 antibodies per individual, 6×10^9 individuals, and many possible strains of the disease or antigen. The other reason is that evolution dynamics occurs at the amino acid level. Use of random energy theories to treat correlations in seemingly intractable physical systems harks back to at least Bohr's random matrix theory for nuclear cross-sections [30] and has been used for disordered mesoscopic systems, quantum chromodynamics, quantum chaos, and quantum gravity [31]. Spin glass models were first advanced in the context of magnetic ions randomly distributed in a nonmagnetic alloy [32–34]. The slow dynamics due to the "frustration effect" in spin glass models leads to nontrivial long-time properties and endows spin glass models with fundamental physical importance [35]. In biology, spin glass models have been applied to explain the origin of biological systems, to simulate RNA evolution, and to study protein folding and function [15, 36–40]. The random interactions occurring between biological agents have been captured by spin glass models. Population- and sequence-level biological dynamics have been studied by spin glass models, capturing the degeneracy in sequence space of biological function.

Table 7.1. *Parameters and values for the generalized NK model of antibodies.*

Parameter	Value	Definition
K	4	Local interaction range within a subdomain
M	10	Number of secondary structure subdomains
N	10	Number of amino acids in each subdomain
P	5	Number of amino acids contributing directly to the binding
L	5	Number of subdomain types (e.g., helices, strands, loops, turns, and others)
a_j		Identity of amino acid at sequence position j
α_i	$1 \le \alpha_i \le L$	Type of secondary structure for the ith subdomain
σ_{α_i}		Local interaction coupling within a subdomain, for subdomain type α_i
D	6	Number of interactions between subdomains
σ_{ij}^k		Nonlocal interaction coupling between secondary structures
σ_i		The ith direct coupling between antibody and antigen

Source: Sun et al. (2006)[44].

Topics such as the diversity, stability, and evolvability of biological systems have been studied [18, 41]. Related to the present discussion are random energy models of spin glasses [33, 34], coarse-grained models of protein folding [37, 38], and N/K-type models of evolution [27, 39, 42, 43]. In detail, we use the generalized NK model, which considers three distinct types of interactions in the antigen/antibody complex: interactions within a subdomain (U^{sd}), interactions between subdomains ($U^{\mathrm{sd-sd}}$), and direct binding interactions between an antibody and an antigen (U^{c}). The parameters of the generalized NK model have been calibrated in the context of protein evolution [21, 27, 40, 45, 46]. The parameters and their values are listed in Table 7.1. The energy function of an antibody binding to an antigen is given by

$$U = \sum_{i=1}^{M} U_{\alpha_i}^{\mathrm{sd}} + \sum_{i>j=1}^{M} U_{ij}^{\mathrm{sd-sd}} + \sum_{i=1}^{P} U_i^{\mathrm{c}}, \tag{7.2}$$

where M is the number of antibody secondary structural subdomains, and P is the number of amino acids in the antibody contributing directly to the binding. The subdomain energy U^{sd} is given by

$$U_{\alpha_i}^{\mathrm{sd}} = \frac{1}{\sqrt{M(N-K+1)}} \sum_{j=1}^{N-K+1} \sigma_{\alpha_i}(a_j, a_{j+1}, \dots, a_{j+K-1}), \tag{7.3}$$

where N is the number of amino acids in a subdomain, and $K = 4$ specifies the range of local interactions within a subdomain. A subdomain is a short sequence of amino acids, of the size of a secondary structural element. There are interactions between the neighboring amino acids, which are captured by $U_{\alpha_i}^{sd}$ within all the subdomains. This value of K was fit by comparison to phage display evolutionary experiments [46]. Another reason for a local interaction range of $K = 4$ is that the helical period in an alpha helix is 3.7 amino acids. Each subdomain belongs to one of $L = 5$ different types (e.g., strands, loops, turns, helices, and others). The Gaussian random number σ_{α_i} is quenched and is different for each subdomain type, α_i. We consider the five chemically distinct amino acid classes (e.g., negative, positive, polar, hydrophobic, and other) [47], since in the B cell model each different type of amino acid behaves as an entirely different chemical entity in the random energy model. The argument a_j is in the range 1 to 5, depending on the class of the amino acid. There is a zero mean and unit variance of all of the Gaussian σ values. The interaction energy between secondary structures is

$$U_{ij}^{sd-sd} = \sqrt{\frac{2}{DM(M-1)}} \sum_{k=1}^{D} \sigma_{ij}^k (a_{j_1}^i, \ldots, a_{j_{K/2}}^i; a_{j_{K/2+1}}^j, \ldots, a_{j_K}^j). \tag{7.4}$$

The number of interactions between secondary structures is set to $D = 6$. The term U_{ij}^{sd-sd} captures nonlocal interactions between amino acids. A typical amino acid has roughly 12 interacting neighbors (in the dense fluid of amino acids in the protein), and half of these, $[2(K-1)]$, are in U^{sd}, and the other half, $[D(M-1)/N]$, are in U^{sd-sd}. The values of σ_{ij}^k and the interacting amino acids, $j_1, \ldots, j_K$, are chosen randomly for each interaction (k, i, j). The values of σ_{ij}^k are quenched Gaussian random numbers with zero mean and unit variance. The chemical energy of binding between amino acids of the antibody and the antigen is given by

$$U_i^c = \sigma_i(a_i)/\sqrt{P}. \tag{7.5}$$

The amino acid that is interaction, i, and the weight of the binding, σ_i, also a Gaussian with unit variance and zero mean, are chosen at random. Using experimental results, we take a typical size of $P = 5$ amino acids contributing directly to the binding.

The generalized NK model, while a simplified description of real proteins, captures much of the thermodynamics of protein folding and ligand binding. In the model, a specific B cell repertoire is represented by a specific set of amino acid sequences. Moreover, a specific instance of the random parameters, i.e., set of σ values, within the model represents a specific antigen. Different antigens correspond to different sets of σ parameters. Two antigens that differ in sequence by fraction p correspond to two sets of σ parameters that differ by p as well. An immune response that finds a B cell that produces an antibody with a high affinity constant to a specific antigen corresponds in the model to finding a sequence having a low energy for a specific parameter set.

The variable region of the light and the heavy chain of an antibody is roughly 100 amino acids long. Most of the binding typically occurs in the heavy chain. We, therefore, choose to focus on a sequence length of 100 residues, representing the heavy chain, in our model of the antibody evolution [48, 49]. Since there are approximately ten secondary structures in a typical antibody, we choose $M = 10$ and thus choose $N = 10$. The human immune system contains roughly 10^8 B cells of different specificities, and the probability that a specific B cell recognizes a particular antigen is approximately 1 in 10^5 [20]. We therefore use 10^3 sequences during the immune response.

A hierarchical strategy is employed in the immune system to search sequence space for high affinity antibodies. In the model, the VDJ recombination process is mimicked by combining optimized subdomains, and the somatic hypermutation process is mimicked by point mutation and selection [21]. The natural process of evolution within the immune system is shown by the solid lines in Fig 7.1. The naive B cell repertoire, which is created in the immune system by combinatorial joining of gene segments, is modeled by choosing subdomain sequences from pools with N_{pool} amino acid segments that were obtained by minimizing U^{sd} for each type of secondary structure. We choose $N_{pool} = 3$ sequences from the top 300 sequences of each subdomain type to reproduce the potential heavy-chain diversity of 3×10^{11} in the immune system [20], because $(N_{pool} \times L)^M \cong 6 \times 10^{11}$.

The local optimization of these starting antibody sequences produced by VDJ recombination occurs by somatic hypermutation. The rate of somatic hypermutation is roughly one mutation per variable region of light and heavy chain per cell division, and B cells divide every 6 to 8 hours during this process [50]. Thus, in our simulation, we perform 0.5 point mutations per heavy-chain sequence. We keep the highest affinity $x = 20\%$ sequences. We amplify these top sequences back to a total of 10^3 copies in one round, all in roughly 1/3 day. The probability of picking the top sequences, which may be mutated, for the next round is $p_{select} = 1/200$ for $U \leq U_{200}$ and $p_{select} = 0$ for $U > U_{200}$, where U_{200} is the 200th best energy out of the energy of all 10^3 sequences after the mutations. This equation is employed 10^3 times for the stochastic process of selecting the 10^3 sequences for the next round. For a specific antigen, represented by a specific set of σ interaction parameters, 30 rounds (10 days) of point mutation and selection are conducted in one immune response. By this means, memory B cells specific for the antigen are created.

We calculate the binding constant, (7.1), as a function of the interaction energy by

$$K^{eq} = \exp(a - b\langle U \rangle). \qquad (7.6)$$

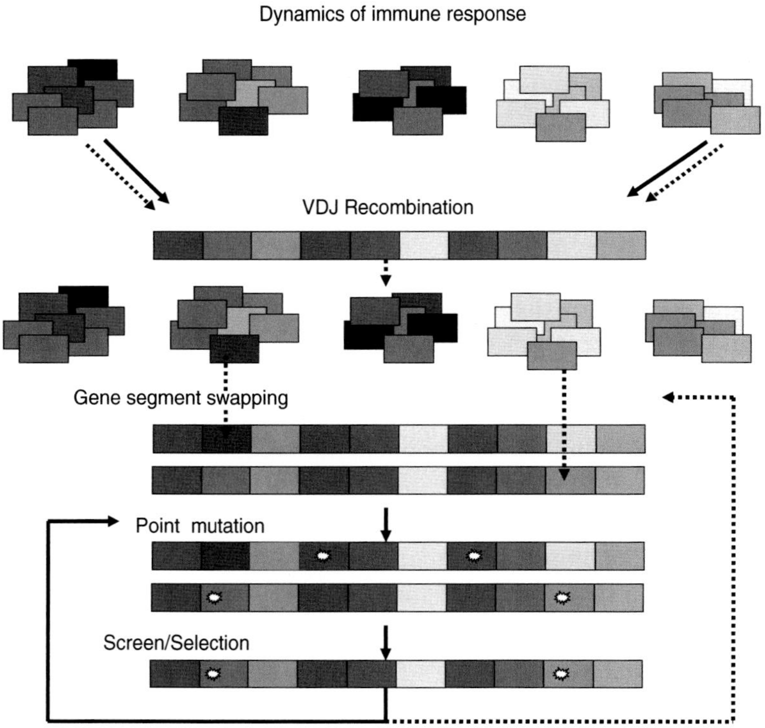

Dynamics of immune response

Fig. 7.1 The dynamics of the immune system in searching the sequence space for high affinity antibody is drawn. The flow described by solid lines is the dynamics used by the natural immune system. The flow represented by dashed lines is the dynamics that could be utilized by the immune system. (Reprinted with permission from [44]. Copyright (2006), World Scientific Publishing.)

Here the constants a and b will be determined by the dynamics of the selection and mutation process. The value $\langle U \rangle$ is the averaged best energy over different instances of the random σ. Typical binding constants of naive antibodies immediately after VDJ recombination are 10^4 l/mol, immediately after the first response of somatic hypermutation are 10^6 l/mol, and immediately after a second response of somatic hypermutation are 10^7 l/mol [20]. These numbers imply that we should take $a = -18.56, b = 1.67$, and $x = 20\%$.

T Cell Model

There are four different kinds of interactions in the generalized NK model of the T cell response, since TCRs bind to peptides presented on MHC molecules. There are interactions within a subdomain of the TCR (U^{sd}), interactions between subdomains of the TCR ($U^{\text{sd-sd}}$), interactions between the TCR and the peptide ($U^{\text{pep-sd}}$), and direct binding interaction between the TCR and peptide (U^c) [21, 24, 51]. The direct interactions come from a limited number of "hot spot" amino acid interactions. The model gives the free energy (U) as a function of the TCR (a_j) and epitope (a_j^{pep}) amino acid sequence. Briefly:

$$U = \sum_{i=1}^{M} U_{\alpha_i}^{sd} + \sum_{i>j=1}^{M} U_{ij}^{sd-sd} + \sum_{i=1}^{M} U_i^{pep-sd} + \sum_{i=1}^{N_b} \sum_{j=1}^{N_{CON}} U_{ij}^c. \qquad (7.7)$$

$$U_i^{pep-sd} = \sqrt{\frac{1}{DM}} \times \sum_{k=1}^{D} \sigma_i^k (a_{j_1}^{pep}, \ldots, a_{j_{K/2}}^{pep}; a_{j_{K/2+1}}^i, \ldots, a_{j_K}^i). \qquad (7.8)$$

$$U_{ij}^c = \frac{1}{\sqrt{N_b N_{CON}}} \sigma_{ij} (a_{j_1}^{pep}, a_{j_2}). \qquad (7.9)$$

This T cell model is similar to the B cell spin glass model. The subdomain energy U^{sd} is as in (7.3), and the interaction between secondary structure U^{sd-sd} is as in (7.4). Here $M = 6$ is the number of TCR secondary structural subdomains, $N_b = 3$ is the number of hot-spot amino acids that directly bind to the TCR, and $N_{CON} = 3$ is the number of T cell amino acids contributing directly to the binding of each peptide amino acid. The term $N = 9$ is the number of amino acids in a subdomain, and $K = 4$ is the range of local interaction within a subdomain. All subdomains are one of the $L = 5$ different types (e.g., strands, loops, turns, helices, and others). The Gaussian random number σ_{α_i} is quenched and different for each value of its argument for each subdomain type, α_i. All the σ values in the model are Gaussian random numbers with zero mean and unit variance. The σ values are distinct for each value of the argument, superscript, or subscript. The term α_i defines the type of secondary structure for the ith subdomain, $1 \le \alpha_i \le L$. We take the number of interactions between secondary structures to be $D = 2$, since TCRs are a bit smaller than antibodies, although the precise value of D turns out not to matter. The values of σ_{ij}^k and the identity of the interacting amino acids, $j_1, \ldots, j_K$, are selected at random for each interaction (i, j, k). The σ_i^k and the interacting amino acids, $j_1, \ldots, j_K$, are also selected at random in the peptide and TCR subdomain for each interaction (i, k). The contributing amino acids, j_1, j_2, and the weight of the binding, σ_{ij}, also a quenched Gaussian random number with unit variance and zero mean, are chosen at random for each interaction (i, j). There are N_b possible values for j_1, and NM possible values for j_2. We consider all 20 amino acids in the T cell spin glass model, (7.7). To consider the differing effects of nonconservative and conservative mutations, we set the random σ for amino acid i that belongs to group j as $\sigma = w_j + w_i/2$, where the w are Gaussian random numbers with zero average and unit standard deviation. There are 5 groups, with 2 amino acids in the negative and polar group, 3 amino acids in the positive and polar group, 3 amino acids in the

Table 7.2. *Parameter values for the generalized NK model of TCRs*

	Parameter	Value	Definition
TCR	a_j		Identity of amino acid at sequence position j
	M	6	Number of secondary structure subdomains
	N	9	Number of amino acids in each subdomain
	L	5	Number of subdomain types (e.g., helices, strands, loops, turns, and others)
	α_i	$1 \leq \alpha_i \leq L$	Type of secondary structure for the ith subdomain
	K	4	Local interaction range within a subdomain
	σ_{α_i}		Local interaction coupling within a subdomain, for subdomain type α_i
	D	2	Number of interactions between subdomains
	σ_{ij}^k		Nonlocal interaction coupling between secondary structures
Epitope	a_j^{pep}		Identity of amino acid of epitope at sequence position j
	N	9	Number of amino acids in epitope
	V	1–4	Number of epitopes on tumor cell
TCR–epitope	N_b	3	Number of hot spot amino acids in the epitope
	N_{CON}	3	Number of amino acids in TCR that each hot spot interacts with
	σ_{ij}		Interaction coupling between TCR and epitope
	σ_i^k		Interaction coupling between TCR secondary structure and epitope
Random couplings	w		Gaussian random number with zero average and unit standard deviation
	σ	$w_j + w_i/2$	Value of coupling for amino acid i of nonconservative type j

Source: Yang *et al.* (2006)[24].

nonpolar with ring group, 4 amino acids in the nonpolar without ring group, and 8 amino acids in the neutral and polar plus cystein group. Table 7.2 lists all the parameters of the generalized NK model.

The binding constant is related to the energy by (7.6). T cells do not evolve; the best T cells are simply selected for. So that the binding constant of the best T cell may fluctuate, we determine the values of a, b in each instance of the ensemble by fixing the geometric average TCR:p-MHC I affinity to be $K = 10^4$ l/mol and minimum affinity to be $K = 10^2$ l/mol [52] for the $V \times N_{size} = V \times 10^8/10^5 = V \times 1000$ naive TCRs that respond to all V epitopes [52, 53]. We are here considering a vaccine against an antigen with V epitopes. This procedure leads to a binding constant of the the highest affinity TCR that fluctuates between 10^5 l/mol and 10^7 l/mol for the different epitopes, in agreement with data [54].

Specific lysis is a standard immunological measure of the probability that an activated T cell will recognize and kill a cell that is expressing a particular peptide-MHC I complex. It is given by [51]:

$$L = \frac{zE/T}{1 + zE/T},\tag{7.10}$$

where E/T is the effector to target ratio. The quantity z is the average clearance probability of one TCR:

$$z = \frac{1}{N_{\text{size}}} \sum_{i=1}^{N_{\text{size}}} \min(1, K_i/10^6).\tag{7.11}$$

Specific lysis correlates well with vaccine efficacy [55] and is a standard measure of the immunological response to a vaccine. TCR affinity is highly correlated with proliferation [56, 57], and on the order of 1–3 TCR/peptide–MHCI interactions are enough for killing [58]. Binding constants larger than approximately 10^6 l/mol do not increase the lysis, which implies the bound in (7.11).

The VDJ recombination process creates the naive TCR repertoire randomly from gene fragments. This is accomplished in the theory by constructing the TCR sequences from the optimized subdomain pools. A TCR sequence is built by randomly determining the type of each of the M subdomains, and then randomly selecting sequences from the subdomain pool of the appropriate type for each subdomain. To construct the pools for each of the subdomain types, 13 of the 100 lowest-energy subdomain sequences are placed. This diversity mimics the known TCR diversity, $(13 \times L)^M \approx 10^{11}$ [59]. Since only 1 in 10^5 naive TCRs responds to a particular antigen with high affinity, and since there are roughly 10^8 distinct TCRs present at any one point in time in the human body [52, 53], the primary response commences with a repertoire of $N_{\text{size}} = 10^3$ distinct TCRs. The initial TCR repertoire is redetermined for each realization of the model, since the U^{sd} that defines the TCR repertoire is different in each instance of the ensemble.

Rounds of cell division, which leads to concentration expansion, and selection for improved binding constants occur in the T-cell-mediated response. The primary T cell response increases the concentration of TCRs that recognize antigen by 1000-fold over ten rounds, and those T cells double approximately every 12–24 hours [20]. The diversity of memory sequences is believed to be 0.5% of that of the naive repertoire [60]. Primary response selection that leads to this diversity is performed through ten rounds, with the top $x = 58\%$ sequences chosen at each round. These ten T cell divisions achieve the concentration expansion by a factor of 10^3 in the primary response, since $2^{10} \approx 10^3$. This process also leads to 0.5% diversity of the memory repertoire, since $0.58^{10} \approx 0.5\%$. In vivo T cell tracking experiments have shown that T cells interact with antigen in lymph nodes up to 5 days after initial stimulation [61, 62], and studies of draining lymph nodes have found that there is antigen capable of priming naive T cells present up to 7 days post inoculation [63]. Both these observations support the idea that the selection process may be operative during each of the ten T cell divisions. Although the precise mechanism of T cell expansion during the primary immune response remains a bit elusive, it is true that the expansion of T cells is nonlinear [64], and that there is typically competition among the T cells for presented antigen [57]. [1]

For the secondary response, if memory TCRs are used, the top $x = 58\%$ of the sequences are chosen, and three rounds of selection are performed [64, 65]. This mimics the concentration increase of approximately $10 \approx 2^3$ that occurs during the secondary memory response [20]. Alternatively, if the naive TCRs are used in the secondary response, the selection and expansion process is identical to that of the primary response. The whole secondary response is a combination of these two, starting with a mixture of naive and memory TCRs. As in ref. [21], we take the fraction of memory cells participating in the secondary response to be proportional to the average binding constant of the memory TCRs with the epitope, which is proportional to Z_m, the clearance probability of the memory sequences. Similarly, we take the fraction of naive TCRs participating in the secondary response to be proportional to the average binding constant of the naive TCRs with the epitope, which is proportional to the clearance probability, Z_n, of the naive sequences.

When considering different epitopes, which differ by

$$p_{\text{epitope}} = \frac{(\text{nonconservative} + \frac{1}{2}\text{conservative}) \text{ amino acid differences in epitope}}{\text{total number of amino acids in epitope}},$$

(7.12)

the random parameters of the generalized NK model and the VDJ selection pools differ by p_{epitope} [21]. To average over a population of people with different immune systems, an average over many instances of these random epitope sequences, models, and VDJ selection pools differing by p_{epitope} is taken.

Below we will discus dengue fever, for which there are four distinct serotypes, each with a different epitope. For dengue, the NS3 nonstructural protein is

---

[1] It is not clear whether this competition occurs during each and every round of T cell division during the primary response. However, the general results of our model hold for any type of exponential expansion during the primary response of the T cells that best recognize the epitope.

an attractive vaccine target [66]. The four epitopes differ by roughly a single conservative amino acid change, [67] and so $p_{\text{epitope}} = 0.5/9$. Thus, in the theory, we choose the four different dengue virus epitopes so that they differ by the requisite p_{epitope}. To average over instances of the model, four new random epitopes were generated each time. The initial TCR repertoire was redetermined for each realization of the model, since the U^{sd} that defines the TCR repertoire is different in each instance of the ensemble. We will also introduce the parameter *mixing round*. In rounds 1 to *mixing round* -1, the four different epitopes are in different lymph nodes, and so the T cells are independently selected for each epitope, and in rounds *mixing round* to 10, the epitopes are assumed systemically mixed, and so the T cells compete for activation from all four of the epitopes. Physiologically, we expect *mixing round* to be in the range 6 to 10.

7.4   Applications and predictions

Original antigenic sin

An immune system response to antigen leads to the establishment of memory against that antigen [68]. Immunological memory endows the immune system with the ability to respond rapidly and effectively to antigens previously encountered. Specific memory is maintained in the DNA of long-lived memory B cells, which persist for many years without residual antigen [69, 70]. Although the immune system is highly effective, some limitations have been observed. The phenomenon of "original antigenic sin" was first noticed as the tendency for antibodies produced during response to first exposure to influenza virus antigens to suppress creation of new and different antibodies in response to different versions of the flu [71, 72]. Roughly speaking, the immune system responds only to the antigen fragments, or epitopes, that are in common with the original flu virus, but recognition of these partially shared epitopes may not lead to full immune clearance. One result is that individuals vaccinated against one strain of the flu may become *more* susceptible to infection by mutated strains of the flu, in comparison to individuals without vaccination – a feature of vaccination we would like to understand and to engineer to be a low-probability event. The mechanism of how original antigenic sin works, even at a qualitative level, is poorly understood.

We describe the dynamics of antibody affinity maturation as a search in antibody sequence space for improved binding energy between antibody and antigen. There are two types of suboptimal dynamics that may lead to original antigenic sin. One mechanism, shown in Fig 7.2a, is localization in sequence space. The antibody sequence corresponding to the best antibody for this year's flu virus (original antigen), may correspond to a local minimum in the sequence

(a)   (b)

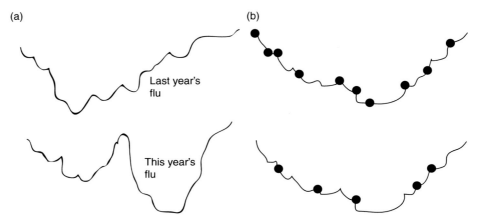

Fig. 7.2 (a) Localization in sequence space in the left plot. The original energy landscape (top) corresponds to immune recognition of the first antigen. The immune recognition of a second antigen corresponds to a new energy landscape (bottom). The barrier between the antibody sequences favorable to the old and new antigens causes localization of antibodies induced to the first antigen during the response to the second antigen. (b) Reduction of diversity of the antibody repertoire, represented by the black dots in the right plot. More diversity on the top landscape and less on the bottom. The memory mechanism puts a restriction on the diversity and leads to original antigenic sin. (Reprinted with permission from ref. [44]. Copyright (2006), World Scientific Publishing.)

space of antibodies against next year's flu virus (new antigen). This phenomenon may occur because the best antibodies that recognize the original antigen are different from those that recognize the new antigen. For the antibodies to evolve to the new global minimum, the barrier between the new global minimum and the old global minimum, now a local minimum, must be overcome. The width and height of the barrier determines the extent of the localization, with stronger localization leading to slower dynamics and increased likelihood of original antigenic sin. The second mechanism is related to the diversity of the initial antibody sequences available during the first response versus those available in the second response, shown in Fig 7.2b. A greater diversity, or number of starting points, leads to a better chance of finding more favorable sequences during somatic hypermutation. Each starting sequence finds only a local minimum, since the landscape upon which the local diffusion occurs is rugged. We will show that the memory sequences produced by the initial exposure reduce the diversity of the secondary response. Thus, the localized sequences produced during the first response reduce the immune system's ability to subsequently respond to related, but different, antigens. It is this competition between memory B cell sequences evolved by somatic hypermutation and naive B cell sequences generated from VDJ recombination that is responsible for original antigenic sin in the immune system.

The number of memory and naive B cells participating in a secondary immune response is estimated by the ratio of the respective binding constants. From the definition of the binding constant, $K^{eq} = $ [Antigen : Antibody] /{[Antigen] [Antibody]}, the probability of binding is proportional to the concentration of antigen-specific antibody, which is 10^2 times greater for the memory sequences [20], and to the binding constant. The average affinity of the 10^3 memory cells preexisting in the repertoire, K_m^{eq}, and of the 10^3 naive B cells in the repertoire, K_n^{eq}, is measured. The ratio $10^2 K_m^{eq}/K_n^{eq}$ gives the fraction of memory cells to naive cells employed in the secondary response. For the secondary response, we perform 30 rounds (10 days) of point mutation and selection, starting with $10^5 K_m^{eq}/(10^2 K_m^{eq} + K_n^{eq})$ memory cells and $10^3 K_n^{eq}/(10^2 K_m^{eq} + K_n^{eq})$ naive cells [73]. The secondary dynamics may lead to better or worse affinities, depending on how similar the first and second antigen are, which is determined by the fraction of amino acids in the epitope that are different, p.

Figure 7.3 shows the evolved binding constant after the secondary response for the case of prior exposure to antigen (solid line) or no prior exposure (dashed line) as a function of the difference, or "antigenic distance" [74], between the first and second antigen, p. The antigenic distance, p, is given by the probability

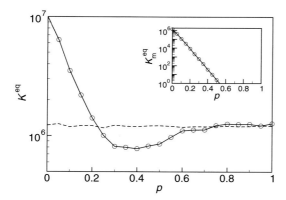

Fig. 7.3 The evolved affinity constant to a second antigen after exposure to an original antigen that differs by probability p (solid line). The dotted line represents the affinity constant without previous exposure. The affinity constant is generated by exponentiating, as in (7.6), the average of the best binding energy, using 5000 instances of the model. In inset is shown the affinity of the memory sequences for the mutated antigen. [Reprinted with permission from ref. [21]. Copyright (2003), American Physical Society.]

of changing each of the parameters of interaction in the subdomain (U^{sd}), subdomain–subdomain (U^{sd-sd}), and chemical binding (U^c) terms of the generalized NK model. Within U^{sd}, we change only the subdomain type, α_i, not the parameters σ_α, which are determined by structural biology and are independent of the antigen. All of the other parameters change, not just those in U^c, because when the antigen changes, the antibody sequences that recognize the antigen change as well. When the antigenic distance is small, exposure to a first antigen leads to a higher affinity constant during the secondary response than without exposure, which is why vaccination and immune system memory is normally effective. For a large antigenic distance, the memory B cells evoked by the antigen previously encountered are uncorrelated with those produced during the second exposure, so immune system memory does not play a role. Interestingly, for intermediate antigenic distances, the immunological memory from the first exposure leads to worse protection, i.e., a lower affinity constant, than does absence of memory – which is the original antigenic sin phenomenon.

We notice that the binding constant, K, decreases exponentially with antigenic distance, as shown by the inset of Fig. 7.3. This exponential decay is a characteristic feature of our model and a characteristic feature of nature. Alternatively, we notice that the binding energy, U, increases linearly with the antigenic distance, p. The linear increase of the binding energy with p stems from the random change of each of the interaction parameters with probability p. The fraction p of new, changed σ values in U^{sd-sd} and U^c are centered around zero and negligible compared to the original, negative evolved values. In U^{sd}, also a fraction p of the α_i are changed. Each changed subdomain will with probability $1/5$ be the same type, otherwise it is different. The original, evolved sequence of the antibody gives a value of roughly zero in an uncorrelated, changed subdomain type α_j. The overall energy lost in U is, therefore, proportional to p. As a detail, the initial energy when $p = 1$ is $1/5$ times the initial energy when $p = 0$, or, $U(p = 1, T = 0) = U(p = 0, T = 0)/5$, where T is time. This relation results because at $p = 1$ the only correlations are that α_i remains unchanged 20% of the time.

A new order parameter to describe antigenic distance

To relate the theory of Fig 7.3 quantitatively to influenza vaccine efficacy, we must determine exactly how the parameter p is related to antigenic distance. Crystallographic data and immunoassays show that only the five epitope regions on the surface of the hemagglutinin protein of influenza A are significantly involved in neutralizing immune recognition [75]. Thus, we were naturally motivated to make the novel assumption that p is related only to differences in these five epitope regions. We introduce $p_{epitope}$, where

$$p_{epitope} = \frac{\text{number of amino acid differences in the dominant epitope}}{\text{total number of amino acids in the dominant epitope}}.$$

(7.13)

The dominant epitope is defined to be the epitope that induces the most significant antibody immune response. The dominant can be measured experimentally by competitive binding assays, although this is not currently done for influenza on a national or worldwide basis. For comparison to epidemiological data, we *define* the dominant epitope of a particular circulating virus strain as the epitope with the largest fractional change in amino acid sequence relative to the vaccine strain for that particular year, [13, 76–79] and we see how well such a definition works. Since our theory models the immune response unfettered by immunosenescence, we limited consideration to epidemiological studies of vaccine efficacy for 18- to 64-year-old healthy subjects in all years since sequencing began, during the years when the H3N2 subtype of influenza A was predominant, and where epidemiological data on vaccine efficacy existed in the literature [13]. We focused on the H3N2 subtype because it is the most common subtype, is responsible for significant morbidity and mortality in humans, and has some available crystallographic, genetic, and epidemiological data. The approach, however, is general.

We test the usefulness of the definition $p_{epitope}$ and the theory of Fig 7.3 on all available literature epidemiological data for H3N2 vaccine efficacy in people [13]. To apply the theory to a candidate vaccine and circulating strain, the identity and sequence of the dominant epitope must be known. The identity and sequence of the vaccine and circulating strains for each year were taken from ref. [81]. The definition of the five epitopes, the surface regions recognized by human antibodies, in the H3N2 hemagglutinin protein were taken from ref. [81]. We determine the vaccine efficacy, E, from literature epidemiological studies [81–95]. Here $E = (u - v)/u$, where u and v are the influenza-like illness rates of unvaccinated and vaccinated individuals, respectively. If the vaccine provides perfect protection against influenza in one year, $E = 1.0$ for that year. A vaccine that is simply useless leads to $E = 0$ for that year. An annual vaccine for which individuals who received the vaccine were more susceptible to influenza than those who did not, leads to $E < 0$ for that year.

The difference between a circulating strain (i.e., original antigenic sin) and a vaccine strain is defined in the model by $p_{epitope}$ (7.13). The vaccine efficacy, E, is assumed to correlate with the binding energy as $E = \alpha \ln[K_{secondary}(p_{epitope})/K_{primary}]$, where the constant α is chosen so that a perfect match between the circulating strain and vaccine gives the average historical

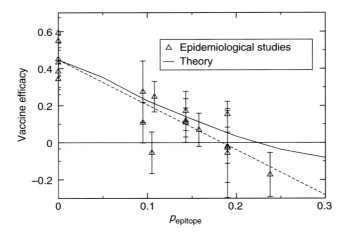

Fig. 7.4 Vaccine efficacy for influenza-like illness as a function of p_{epitope} as observed in epidemiological studies and as predicted by our theory. Also shown is a linear least-squares fit to the data (long dashed, $R^2 = 0.81$). (Reprinted with permission from ref. [13]. Copyright (2006), Elsevier.)

value of $E = 45\%$ vaccine efficacy. The binding constant K_{primary} is the result from the primary immune response, and $K_{\text{secondary}}$ is the result from the secondary immune response following vaccination. The theory is entirely predictive, without fitted parameters, except for the determined constant α. The point at which vaccine efficacy becomes negative, for example, is independent of the value of α.

Figure 7.4 displays the epidemiologically determined vaccine efficacy and the efficacy predicted as a function of p_{epitope} by the theory. The statistical mechanical generalized NK model captures the essential biology of the immune response to influenza virus and vaccination. The results demonstrate the value of using p_{epitope} to define antigenic distance, or degree of antigenic drift. When the antigenic distance, p_{epitope}, in the dominant epitope is greater than 0.19, by historical epidemiological data, or 0.22, by theory, the vaccine efficacy becomes negative (see Fig. 7.4). Error bars are calculated assuming binomial statistics for each data set: $\varepsilon^2 = [\sigma_v^2/u^2/N_v + (v/u^2)^2\sigma_u^2/N_u]$, where $\sigma_v^2 = v(1-v)$ and $\sigma_u^2 = u(1-u)$. If two sets of data are averaged in one year, then $\varepsilon^2 = \varepsilon_1^2/4 + \varepsilon_2^2/4$.

Methods currently employed by the World Health Organization (and the USA Centers for Disease Control) to quantify antigenic distance include calculating the sequence difference in the entire hemagglutinin protein,

$$p_{\text{sequence}} = \frac{\text{number of amino acid differences in the sequence}}{\text{total number of amino acids in the sequence}}, \qquad (7.14)$$

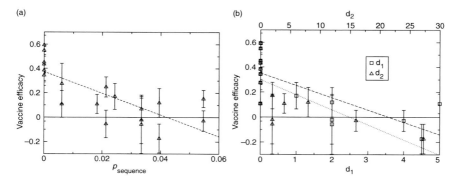

Fig. 7.5 (a) Vaccine efficacy as observed in epidemiological studies for influenza-like illness as a function of $p_{sequence}$ (see 7.14). Also shown is a linear least-squares fit to the data (long dashed, $R^2 = 0.59$). The epidemiological data shown in this figure are the same as in Fig 7.4. Only the definition of the x-axis is different. (b) Vaccine efficacy for influenza-like illness as a function of two measures of antigenic distance, d_1 [74] and d_2 [96], derived from ferret antisera experiments. Experimental data were collected from a variety of sources [74, 97–103]. Results were averaged when multiple hemagglutination inhibition (HI) studies had been performed for a given year. These HI binding assays measure the ability of ferret antisera to block the agglutination of red blood cells by influenza viruses. Also shown are linear least-squares fits to the d_1 (long dashed, $R^2 = 0.57$) and d_2 (short dashed, $R^2 = 0.43$) data. The epidemiological data shown in this figure are the same as in Fig 7.4. Only the definition of the x-axis is different. (Reprinted with permission from ref. [13]. Copyright (2006), Elsevier.)

and ferret antisera inhibition assays [74, 96]. Upon plotting vaccine efficacy against the currently used measures of antigenic distance (see Fig 7.5a, b), one sees that $p_{epitope}$ correlates to a greater degree with epidemiological human efficacy than do the existing methods, even ferret animal model studies. It is perhaps not overly surprising that the ferret animal model provides only an approximation of the human immune response.

It is logical to expect to improve the predictive ability of the order parameter $p_{epitope}$ through the inclusion of additional variables in the measure. We have tested a number of potential improvements to the original definition of the $p_{epitope}$ order parameter [44]. For example, we included the influence of subdominant epitopes. We made a distinction between conservative and nonconservative amino acid differences. We considered amino acid differences at sites adjacent to the epitope regions. We considered amino acid antigenic drift in the neuraminidase protein. None of these new measures had a significantly better correlation with the literature epidemiological efficacy data available to us to date. It is possible that with additional human vaccine efficacy data, some of these refinements will provide an improvement over p_{epitope}, although the original definition is relatively successful to date.

Glassy dynamics suppresses autoimmune disease

Autoimmune disease occurs when an immune response is directed by mistake against one's own proteins [20]. In most cases, the bind of an antibody or TCR against a foreign antigen is highly specific. Sometimes, however, the antibody or TCR involved can bind other antigens, and this phenomenon is termed cross-reactivity [104]. Cross-reactivity occurs when the unrelated antigen has chemical features in common with the original antigen. By measuring the affinity of the antibody for the other antigen, cross-reactivity is quantified experimentally. Cross-reactivity is one of the mechanisms by which autoimmune disease may develop. Mechanisms of tolerance at several biological levels usually act to reduce the possibility of autoimmune disease. Both the environment and genetic makeup influence an individual's susceptibility to autoimmune disease.

We discuss two results in this section [22]. First, we show how the dynamics of an evolving population of antibodies is affected by different evolutionary mechanisms for changing the sequence. Specifically, the evolutionary dynamics of antibodies created in response to pathogen invasion is studied. Effective dynamics is crucial to the efficacy of the immune response, since the immune system must mount a timely response to the threat. The hierarchical structure of the generalized NK model plays a critical role in the evolutionary dynamics. Second, we show that it is possible with a biologically plausible evolution process to find antibodies with significantly higher affinities for antigen than those produced by the normal primary immune response. As we will show, however, these antibodies are more cross-reactive and would greatly increase susceptibility to autoimmune disorders than do natural antibodies. The biological concept of cross-reactivity is related to the physical concept of the chaos exponent in spin glasses [105, 106], except that we are interested in the immediate energy response for cross-reactivity and the equilibrated response to a change in the couplings for the chaos exponent. Taking our two results together, we suggest that a careful balance has evolved between binding affinity to and specificity for foreign antigen in the primary adaptive immune response.

We use two different theoretical strategies to search protein sequence space for high-affinity antibodies. The first strategy is the model of the normal B cell immune response that we have already discussed. It starts with combinatorial joining of optimized subdomains, followed by sequences undergoing rounds of point mutation (PM) and selection. This procedure is shown by the solid lines in Fig. 7.1. This is a model of the VDJ recombination, somatic hypermutation, and clonal selection that occurs in normal B cell development [104]. We generate five optimized subdomain pools, each composed of the 300 lowest-energy subdomains, corresponding to the $L = 5$ types. We randomly choose three sequences from each subdomain pool as our pool of material for possible VDJ recombinations, mimicking the known potential antibody diversity, $(3 \times 5)^{10} \approx 10^{11}$ [21].

In our second strategy, we include a more powerful gene rearrangement process during the whole process of antibody evolution. As in the normal immune response, VDJ recombination is used to generate the initial population. Then during each round of evolution and selection we perform gene segment swapping (GSS) in addition to somatic hypermutation, as shown by the dashed lines in Fig 7.1. GSS-type processes are used experimentally to produce antibodies with binding constants $\approx 10^{11} - 10^{13}$ l/mol [26], and these processes exist within the biological hierarchy of evolutionary events [16–19].

In GSS, the sequence of a subdomain of the five types is replaced by another sequence from the optimized subdomain pool of the same type. In both strategies, each sequence undergoes an average of 0.5 point mutations per round of selection. In GSS + PM, there is a probability of 0.05 of each subdomain in a sequence being replaced by GSS. After the mutation, selection occurs, and the 20% highest-affinity antibodies are kept and amplified to form 10^3 sequences for the next round of mutation and selection. Thirty rounds of affinity maturation occur in the primary response [21], during which B cells undergo clonal selection and differentiate into plasma cells and memory cells in response to foreign antigen. The presented results are averaged over 5000 instances of the ensemble.

The average affinity of the antibodies improves during the primary response due to mutation and selection. The evolution of the binding energy as it occurs due to the two different strategies is shown in Fig. 7.6. The GSS + PM dynamics yields sequences with lower energies than does the PM dynamics. In other words, GSS+PM is more effective than PM in searching sequence space for antibodies of higher affinity for a given antigen. The best binding energy produced by the GSS + PM dynamics during 30 rounds of primary response, averaged over 5000 instances of the ensemble, is -21.9 compared by a value produced by the PM dynamics of -19.7. These energies corresponding to affinities of $K = 6.7 \times 10^7$ l/mol and $K = 1.6 \times 10^6$ l/mol, respectively. So, GSS+PM evolves the affinity an order of magnitude more than does PM during the primary response, which is even better than the PM dynamics does during the secondary response [21]. That is, the correlated movement of multiple amino acids that occurs in GSS+PM accelerates the optimization of $U^{\text{sd}-\text{sd}}$. Antibodies with higher affinity

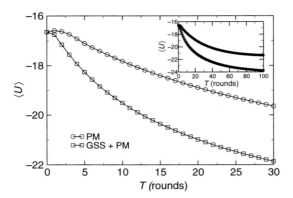

Fig. 7.6 Evolution of the affinity energy for the cases of point mutation (PM) only and gene segment swapping (GSS) plus PM as a function of the number of rounds of mutation and selection used to evolve the population of antibodies. [Reprinted with permission from ref. [22]. Copyright (2005), American Physical Society.]

for the antigen work more effectively in many ways. For example they may neutralize bacterial toxins, inhibit the infectiousness of viruses, or block the adhesion of bacteria to host cells at lower concentrations than would lower affinity antibodies [104]. Based solely on Fig. 7.6, it is difficult to understand why Darwinian evolution did not lead to use of GSS+PM or any other more efficient strategy as the preferred strategy for B cell expansion during the primary response rather than somatic hypermutation.

To begin to answer this evolutionary immunological paradox, we present a calculation of cross-reactivity, thereby quantifying the specificity of the antibodies generated by the two different dynamics. Thirty rounds of primary response dynamics are conducted for both PM and GSS + PM, starting from VDJ recombinants. The cross-reactivity of these antibodies is then measured. That is, the antigen is changed by fraction p, which means each interaction parameter in the generalized NK model is changed with probability p. Affinity constants for the new, modified antigen are calculated in the two cases. It is commonly assumed that if the binding affinity is less than 10^2 l/mol, the binding is nonspecific. We therefore use $K > K_c = 10^2$ l/mol to determine in each case at what value of p the cross-reactivity stops. As shown in Fig. 7.7a, the cross-reactivity ceases at larger p in the GSS + PM case, at $p_2 = 0.472$, than in the PM only case, at $p_1 = 0.368$. Cross-reactivity ceases at a value approximately $\Delta p = 0.10$ larger in the GSS + PM case. In the region of specific binding, $p < 0.472$, the affinity is always better in the GSS + PM case, $K > 10^2$ l/mol. These cross-reactivity measurements show that antibodies generated by GSS+PM can recognize a greater diversity of antigens and with higher affinity than can antibodies generated by the PM dynamics alone. Such cross-reactivity has been experimentally observed

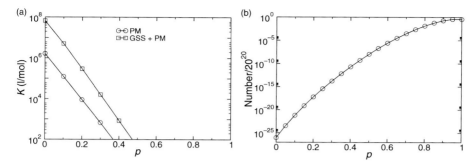

Fig. 7.7 (a) Affinity of memory antibody sequences after a primary immune response for the two different immune system strategies (PM and GSS + PM) to altered antigens. The binding constant is K, and the antigenic distance of the new altered antigen from the original antigen is p. Cross-reactivity ceases at larger distances in the GSS + PM case ($p > p_2 = 0.472$) than in the PM only case ($p > p_1 = 0.368$). (b) The number of possible epitopes that are at an antigenic distance of p from another epitope. The epitopes are assumed to be 20 amino acids in length. The plot is normalized, so that $\sum_p N(p) = 20^{20}$. (Reprinted with permission from ref. [22]. Copyright (2005), American Physical Society.)

in high-affinity TCRs [107]. For the same reason discussed in the inset of Fig. 7.3, the value of K decreases exponentially with the degree of antigenic change, p.

Now a question is how many protein molecules are there between $p_1 = 0.368$ and $p_2 = 0.472$ that can be specifically bound by the antibodies produced through VDJ recombination and GSS + PM, but not by antibodies produced through VDJ recombination and PM only? Antibodies bind to the epitope regions of an antigen, and so we want to calculate how many epitopes these two different classes of antibodies may recognize. We will show that the antibodies produced by GSS + PM recognize 10^3 times more epitopes than do the antibodies produced by the normal immune response. We take a typical epitope length of $B = 20$ amino acids [104, 108]. The total number of possible epitopes is 20^B, since there are 20 different amino acids [109]. We show the normalized number of epitopes at an antigenic distance of p from the natural epitope of an antibody in Fig. 7.7b. The result is given by

$$N(p) = 19^i B!/[i!(B-i)!], \tag{7.15}$$

where $i = pB$. This number is $N(p_1)/20^B = 2 \times 10^{-12}$ and $N(p_2)/20^B = 2 \times 10^{-9}$. The number of epitopes between $p = 0$ and p_1 is approximately $A_1 \approx 2 \times 10^{-12}$, and the number of epitopes between p_1 and p_2 is approximately $A_2 \approx 2 \times 10^{-9}$, since $N(p)$ grows exponentially with p. Using Stirling's approximation since B is large, we find

$$\frac{N(p_2)}{N(p_1)} \sim \exp\{-B\{\ln[(1-p_2)/(1-p_1)] + p_1 \ln[19(1/p_1 - 1)] - p_2 \ln[19(1/p_2 - 1)]\}\}. \tag{7.16}$$

This ratio for 20 amino acids is 10^3. The ratio varies between 160 and 4900 as the epitope size is varied between the known limits of 15–25 amino acids. The ratio varies by 30% with $\pm 10\%$ variation of either p values; the generalized NK model p values agree with experiment to within $\pm 10\%$ [13, 21].

To answer why Darwinian selection has favored the relatively slower dynamics for the immune response, we consider the overall function of the immune system rather than just the binding of a single antibody to a single foreign antigen. There is a delicate balance in the immune system between a weak or no response to self antigens and a strong response to invading nonself antigens. Thus, it is crucial for antibodies produced by the immune system to discriminate one's own proteins from foreign antigens [110–112]. An immune system incapable of recognizing the antigens associated with an invading pathogen and initiating a response would be an inadequate defense mechanism. Conversely, production of antibodies that bind to one's own proteins results in autoimmune diseases, e.g., diseases such as rheumatoid arthritis and type I diabetes. [20] One might think that GSS + PM would be the favored dynamics, if only the immune system were able to stop after ≈ 10 rather than 30 rounds. However, subsequent exposures to related antigen would lead to additional ≈ 10 rounds. After just two such exposures, the GSS + PM dynamics would have evolved an unacceptably large binding constant, saturating after several exposures to approximately 10^{10} l/mol. The PM dynamics, conversely, saturates at approximately 10^7 l/mol. More details of this glassy dynamics can be found in ref. [44]. The saturation of GSS+PM dynamics comes at a lower energy and later time than in PM dynamics. For this reason, GSS + PM is more likely to lead to autoimmunity than is PM.

Cross-reactivity is one mechanism of autoimmune disease [20]. From Fig. 7.7a,b, we see that antibodies produced in the primary immune response recognize a random, 20 amino acid epitope with probability $\approx 10^{-12}$. It is known that a typical cell compartment contains $\approx 10^4$ proteins, each with a length ≈ 500 amino acids [113]. Antibodies recognize only surface epitopes of proteins. The amino acids exposed on protein surfaces are in a loop or turn, with typically 1/3 of the loop or turn exposed. Thus, a typical recognized contiguous fragment is six to seven amino acids long. Given a typical length of 20 amino acids for the entire loop or turn, and the typical epitope size of 20 amino acids, an antibody will recognize approximately three noncontiguous regions of length six to seven amino acids in a protein sequence of 500 amino acids. There are roughly $500^3 \approx 10^8$ such epitopes. Given the roughly 10^4 proteins in a cellular region, there will be $\approx 10^{12}$ total epitopes expressed in each cell. Thus, the number of epitopes recognized by a typical antibody in each cellular compartment is $A_1 \times 10^{12} \approx 1$. Taking a long protein of 1000 amino acids, the number of recognized epitopes increases only to 10. Regulatory mechanisms can handle the occasional aberrant antibody, but cannot handle the 10^3 aberrant antibodies that GSS + PM would produce.

This remarkable result indicates that those antibodies produced by the immune system recognize only their intended target on average. Conversely, antibodies evolved in a biologically possible but hypothetical immune response composed of VDJ recombination followed by a period of GSS + PM recognize on average $A_2 \times 10^{12} \approx 10^3$ epitopes in each cell. Such antibodies, although having higher affinities for the intended target, would lead to 10^3 more instances of autoimmune disease. Such promiscuous antibodies would place too large a regulatory burden on the mechanisms that eliminate the occasional aberrant antibodies [114]. Thus, we suggest that selection has evolved the human immune system so as to generate antibodies recognizing on average only the intended epitope during the B cell immune response. Inclusion of "more efficient" evolutionary dynamics is excluded by the bound $A \times 10^{12} = O(1)$.

It is possible to test experimentally our prediction that proteins evolved with GSS + PM are more cross-reactive than are proteins evolved with PM alone. Techniques such as DNA shuffling [115], exon shuffling [116, 117], and swapping are similar to GSS. In the absence of any additional constraints, we predict that antibodies evolved with these methods would have higher figures of merit [27] as well as more cross-reactivity. That is not to say selection for increased specificity is not possible in protein evolution, but rather that increased antibody affinity leads to increased cross-reactivity in the absence of selective pressures against promiscuity.

Our model predicts that chronic infection may lead to autoimmune disease, even with PM dynamics [44]. Chronic infection is postulated to be responsible for some fraction of autoimmune diseases [118], including rheumatic diseases such as arthritis [119]. However, the significance and strength of this correlation are controversial [120]. Our model suggests a broad distribution for the time of onset of autoimmune disease due to chronic infection [44]. A search for long-time correlations between chronic infection and autoimmune disease may help to resolve some of the controversy. It would be interesting to search for the onset time distribution in experiments, which would serve as one test for the existence of glassy dynamics in antibody evolution.

In summary, a mechanism that combines fast dynamics and glassy dynamics has evolved in the immune system to search amino acid sequence space. The rapid creation of initial antibodies by VDJ recombination serves to jump start the immune response. The glassy dynamics in PM serves a functional purpose to inhibit autoimmunity by slowing down antibody evolution at long times.

Dengue fever and polytopic vaccination

There are four serotypes of the dengue virus. Immunization with one serotype is protective against future challenge with the immunizing virus. However, immunity raised after infection by one serotype protects only modestly or even negatively against reinfection by the others [121–123]. In particular, the risk of dengue hemorrhagic fever, from which essentially all the mortality of dengue stems, during a secondary infection of dengue has been observed to rise significantly if it follows a primary infection from a different serotype [10, 121, 124].

This "original antigenic sin" implies that an effective dengue vaccine must induce protective immunity to all four dengue viruses [7, 121, 122, 125]. To date, however, no such successful four-component vaccine has been developed. The immunological epitopes of the four dengue serotypes are related but differ somewhat in sequence. It is believed that the T cell immunological response to each virus is largely, although not exclusively, to a single epitope [66, 67]. For the epitope-based T cell vaccines discussed here, the response is directly solely to epitopes included in the vaccine. Differences in the epitope sequences of the serotypes affect the quality of the T cell response, and simultaneous exposure to all four dengue serotypes reduces the quality of the immune response to some of the serotypes [7, 126].

In an effort to establish a connection between infection of the mononuclear cells that dengue targets and pathophysiologic changes of dengue disease, some recent research has focused on the T cell response to dengue [121, 125–127]. In particular, some work has gone into quantifying the extent of low-avidity cross-reactivity of T cells between the different dengue serotypes, since cross-reactivity is believed possibly to lead to an increased risk of dengue hemorrhagic fever [121, 122, 125]. Clinical trials of a four-component dengue vaccine [126] show an immunodominance effect in the T cell response, in which the immune response is strong against some of the serotypes, but not against the others. Why T cell immunodominance occurs so strongly against dengue remains mysterious. Most studies suggest that the CD8+ TCRs specific for dominant epitopes of one serotype suppress the response to epitopes of other serotypes, due to resource competition, apoptosis, homeostasis, and reduction of viral load [128–130]. Immunization with nonstructural proteins is also viewed as a possible means to avoid the other problem of antibody-dependent enhancement associated with antibody-based dengue vaccines [122]. While for NS1 an antibody response may contribute to protection, for the other nonstructural proteins such as NS3, it is believed that a primarily T cell response will be induced [122]. The degree to which the T cell response will be beneficial or detrimental remains unclear. Quantitative understanding of immunodominance has been hampered by the complex immunological interactions between the TCRs and the epitopes and by the complex selection and competition process among the TCRs. The focus of our work is on the study of induction of T cell-based immunity. While vaccine-induced antibodies may be required in addition to T cells, for immuno-logical protection against the dengue viruses, understanding the T cell response seems to be important for resolving the issues of original antigenic sin, immun-odominance, and dengue hemorrhagic fever associated with dengue, each of which seem to be based in part or whole on the T cell response.

To develop an effective four-component vaccine against dengue, and to mitigate immunodominance, we here explore the possibility of using poly-topic, or multi-site, vaccination to induce an effective T cell immune response against the four dengue serotypes. By multi-site vaccination, we mean more than simply injecting the same vaccine in multiple places, e.g., as used to be done with the rabies vaccine in the abdomen and buttocks. "Multi-site vacci-nation" here means injection of each component of a vaccine to drain to a physiologically distinct lymph node, as shown in Fig. 7.8. We here focus on multi-site vaccination for the dengue serotypes, for which the dominant epi-topes are related. We investigate whether injection of the four epitopes from the four serotypes in different physical locations sculpts a broader TCR response, since TCRs will thereby be selected for in different lymph nodes. We deter-mine whether polytopic vaccination increases recognition of the four dengue serotypes and reduces immunodominance. Administration of two different vac-cines at physically separated sites rather than the same site has been shown to reduce immunodominance in cancer vaccines in mice [132]. Efficient draining to nearby lymph nodes can be enhanced through use of antigen-bearing dendritic cells [133, 134]. We will also discuss whether subdominant epitope priming is an effective strategy to sculpt an increased number of TCRs recognizing the sub-dominant dengue serotypes. Subdominant epitope priming has proved useful in LCMV and cancer experiments in mice [135–137].

We first compare the theory to results from clinical trials. Agreement within experimental error bars is observed [23] (see Table 7.3). The least dominant response gives a specific lysis value of roughly 0.12 that of the most dominant response, and this four-component vaccine is determined not effective enough clinically. This result implies that an effective vaccine must reduce immun-odominance, so that the least dominant response is greater than 0.12 that of the dominant response.

Since the experimental data are limited, we use two different methods to analyze them. In the first method, we take the vaccine epitope from the serotype that corresponds to the stimulating serotype. In other words, for the sample

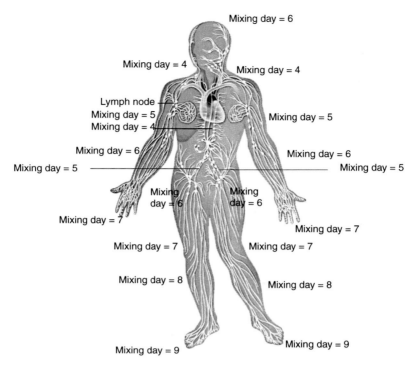

Fig. 7.8 A graphic explanation of polytopic vaccination. Value of the parameter mixing round for vaccination to different lymph nodes at different distances from the heart. Humans have several hundred lymph nodes. For effective polytopic vaccination, well-separated sites on different limbs are used. (Reprinted with kind permission of Springer Science and Business Media from ref. [131].)

that was stimulated with DV1, we take the vaccine epitope from DV1. Since four blood samples from each patient were tested [126], each stimulated by a different serotype, this provides one set per patient and therefore four sets in all. This is the same as ref. [23]. To investigate the limitations posed by the small data set of the first method, we try a different strategy, the second method, where we take all 16 sets of data and calculate a lysis averaged over all four stimulation epitopes. We compare the two methods in Table 7.3.

Due to the limited size of the T cell repertoire and competition for memory expansion to the four dengue serotypes, original antigenic sin may occur in the T cell response. Original antigenic sin does appear to happen, at least in the theory, because the epitopic variation for dengue lies in the range where it is predicted to occur, $0.02 < p_{\text{epitope}} < 0.4$ [23].

An immune response to one or two dominant dengue serotypes suppresses the response to the other, subdominant serotypes. This immunodominance

Table 7.3. *Specific lysis ratios for the least to the most dominant epitope of the four dengue viruses. Experimental data are from ref. [126].*

	Experiment[a]		Simulation
	Method I	Method II	
L_1	0.12 ± 0.03	0.11 ± 0.02	0.12
L_2	0.22 ± 0.02	0.18 ± 0.02	0.24
L_3	0.31 ± 0.04	0.25 ± 0.04	0.33
L_4	0.40 ± 0.09	0.32 ± 0.07	0.46

Source: Zhou and Deem (2006) [23].

[a] We use two methods to analyze the experiment data. In the first method, we take the vaccine epitope from the serotype that corresponds to the stimulating serotype. In the second method we take all 16 sets of data from all stimulating serotypes and calculate p.

Table 7.4. *Clearance probability comparison. The difference between the epitopes of the four dengue viruses is roughly $p_{epitope} = 0.5/9^a$.*

	Normal vaccine[a]		Polytopic vaccine		
$Z1/(Z1 + Z2 + Z3 + Z4)$	0.032	0.253	0.113	0.109	0.097
$Z2/(Z1 + Z2 + Z3 + Z4)$	0.106	0.152	0.285	0.148	0.134
$Z3/(Z1 + Z2 + Z3 + Z4)$	0.256	0.227	0.234	0.377	0.216
$Z4/(Z1 + Z2 + Z3 + Z4)$	0.606	0.368	0.368	0.366	0.554

[a] The normal vaccine column shows the result of a typical four-component dengue vaccine (DV). The polytopic vaccine column shows the result of our new strategy, i.e., a primary single-component DV (virus i in the ith column) is used first, and then a secondary polytopic four-component DV is followed (*mixing round* = 9).

can be seen in Table 7.3. The immunodominance stems from the heterologous nature of the immune system. The immunodominance appears inevitable. Vaccination with only a single virus increases the immunodominance, additionally as a result of original antigenic sin.

Priming with a subdominant epitope first sculpts a broader immune response and leads to reduced immunodominance in the secondary response than does priming with the dominant epitope; compare the first two rows of Table 7.4 to the last two. Epidemiological studies have suggested that the order of exposure to dengue serotypes is important, and that the response is non-commutative [124]. Cytotoxic lymphocytes induced by one dengue serotype may recognize another dengue serotype to a greater extent than the reverse, which increases the odds of dengue hemorrhagic fever [8, 138].

By physically separating the selection of the TCRs and reducing the pressure on resource for stimulation of TCRs within each lymph node, polytopic vaccination can sculpt the TCR repertoire toward the subdominant epitopes, and so reduce immunodominance. We term this the polytopic effect. The existence of immunodominance and competition through space imply that an optimal dengue vaccine should induce a high concentration of high-affinity TCRs to maximize protection, minimize pathologic heterologous immunity, and achieve a long-lasting immune response, against all four dengue virus serotypes [126, 127, 129, 139]. Larger values of the parameter *mixing round* allow for longer periods of independent TCR selection during the typical ten rounds of immune response and lead to less immunodominance. The parameter *mixing round* is relative to the T cell division time, typically 12–24 hours. The value of *mixing round* at lymph nodes different distances from the heart is shown in Fig. 7.8. For the human immune system, the time for T cells to leave the lymph nodes plus the circulation time of the lymph system is in the range 6–10 days [20, 140]. The polytopic effect is positive within this entire physiological range.

By combining polytopic injection with subdominant epitope priming, a vaccination protocol for sculpting the immune response to dengue is achieved. Immunodominance is clear when a typical four-component dengue vaccine is used. We can see immunodominance is largely reduced with the new protocol, as the clearance probability against different dengue virus is nearly the same, since the numbers in the first column of polytopic vaccination are nearly equal. In this case, the response is improved to 4× when the least dominant epitope is used in the primary response, and provides the most immunological benefit.

This new protocol reduces immunodominance more fully than does polytopic injection or subdominant epitope priming alone [23], both of which reduce immunodominance more than does a traditional four-component dengue vaccine. Immunological memory and competition through time imply that response to a subdominant virus can be strengthened with prior exposure.

The polytopic vaccination and subdominant epitope priming protocols appear to apply generically to other, non-dengue, multi-strain viral diseases. Competition between the memory and naive response due to limited partial cross-reactivity is the reason for original antigen sin. Immunodominance is a more general phenomenon, which may persist for all values of $p_{epitope}$, and which also results from competition through space. A reduction of this deleterious competition is why the separate selection that occurs in polytopic vaccination leads to reduced immunodominance and why subdominant epitope priming achieves a more diverse and effective TCR response, with a significant number of TCRs responding to the subdominant epitopes.

7.5   Conclusion

The immune system has evolved to protect the human host against death by infection. We have introduced hierarchical spin glass models of the evolutionary dynamics that occurs in the B cell and T cell immune responses. The model was used to expose the mechanisms for original antigenic sin, wherein an initial exposure to antigen degrades the response of the immune system upon subsequent exposure to related, but different, antigens. The suboptimal evolutionary dynamics leading to original antigenic sin were shown to stem from a reduction in diversity of the antibodies responding to the antigens and from a localization in sequence space of the evolving antibodies. A new order parameter to characterize antigenic distance was introduced. This order parameter correlates with the efficacy of influenza vaccine in humans even better than do results from standard ferret animal model studies used by world health authorities. This new order parameter appears to be a useful new tool for making vaccine-related public health policy decisions. Interestingly, the glassy dynamics of evolution within the immune system was shown to play a functional role through inhibiting autoimmune disease. A balance appears to have evolved in the mechanism for searching amino acid sequence space in the immune system between affinity and specificity. That is, the normal immune response inhibits autoimmune disease at the cost of a weaker average binding affinity. Finally, we have discussed a predictive theory of immunodominance, and applied it to four-component dengue vaccination. From our theory, subdominant epitope priming followed by secondary polytopic vaccination appears to be a promising strategy for dengue fever and other multi-strain diseases.

Acknowledgments
This work was supported by DARPA's FunBio project, with which it is a pleasure to acknowledge stimulating discussions.

References
1   World Health Organization Media Centre Influenza Fact Sheet 211 (2003), www.who.int/mediacentre/factsheets/fs211/en/.
2   J. R. Lave, C. J. Lin, M. J. Fine, and P. Hughes-Cromick, *Sem. Respir. Crit. Care Med.* **20**, 189 (1999).
3   K. M. Neuzil, G. W. Reed, J. E. F. Mitchel, and M. R. Griffin, *J. Am. Med. Assoc.* **281**, 901 (1999).

4   M. J. W. Sprenger, P. G. Mulder, W. E. P. Beyer, R. VanStrik, and N. Masurel, *Int. J. Epidemiol.* **22**, 334 (1993).

5   M. I. Meltzer, N. J. Cox, and K. Fukuda, *Emerg. Infect. Dis.* **5**, 659 (1999).

6   K. Stohr and M. Esveld, *Science* **306**, 2195 (2004).

7   D. J. Gubler, *Clin. Microbiol. Rev.* **11**, 480 (1998).

8   I. Kurane and T. Takasaki, *Rev. Med. Virol.* **11**, 301 (2001).

9   R. V. Gibbons and D. W. Vaughn, *Bio. Med. J.* **324**, 1563 (2002).

10   S. B. Halstead, *Science* **239**, 476 (1988).

11   D. Guha-Sapri and B. Schimmer, *Emer. Themes Epidemiol.* **2**, 1 (2005).

12   H. Varmus, R. Klausner, E. Zerhouni, *et al.*, *Science* **302**, 398 (2003).

13   V. Gupta, D. J. Earl, and M. W. Deem, *Vaccine* **24**, 3881 (2006).

14   Research Report–2004-03-25, Press Releases, American Autoimmune Related Diseases Association, www.aarda.org/press_releases.php.

15   P. W. Anderson, *Proc. Natl Acad. Sci. USA* **80**, 3386 (1983).

16   J. A. Shapiro, *Genetica* **86**, 99 (1992).

17   J. A. Shapiro, *Trends Genet.* **13**, 98 (1997).

18   D. J. Earl and M. W. Deem, *Proc. Natl Acad. Sci. USA* **101**, 11 531 (2004).

19   M. G. Kidwell and D. R. Lisch, *Evolution* **55**, 1 (2001).

20   C. A. Janeway, P. Travers, M. Walport, and M. Shlomchik, *Immunobiology: The Immune System in Health and Disease*, 6th edn. (New York, Taylor & Francis, 2004).

21   M. W. Deem and H.-Y. Lee, *Phys. Rev. Lett.* **91**, 068 101 (2003).

22   J. Sun, D. J. Earl, and M. W. Deem, *Phys. Rev. Lett.* **95**, 148 104 (2005).

23   H. Zhou and M. Deem, *Vaccine* **24**, 2451 (2006).

24   M. Yang, J.-M. Park, and M. W. Deem, *Physica A* **366**, 347 (2006).

25   G. M. Griffiths, C. Berek, M. Kaartinen, and C. Milstein, *Nature* **312**, 271 (1984).

26   R. Schier *et al.*, *J. Mol. Biol.* **263**, 551 (1996).

27   L. D. Bogarad and M. W. Deem, *Proc. Natl Acad. Sci. USA* **96**, 2591 (1999).

28   K. H. Fischer and J. A. Hertz, *Spin Glasses* (New York, Cambridge University Press, 1991).

29   D. J. Gross and M. Mézard, *Nucl. Phys. B* **240**, 431 (1984).

30   N. Bohr, *Nature* **137**, 344 (1936).

31   T. Guhr, A. Müller-Groeling, and H. A. Weidenmüller, *Phys. Rep.* **299**, 189 (1998).

32   S. F. Edwards and P. W. Anderson, *J. Phys. F* **5**, 965 (1975).

33   D. Sherrington and S. Kirkpatrick, *Phys. Rev. Lett.* **35**, 1792 (1975).

34   B. Derrida, *Phys. Rev. Lett.* **45**, 79 (1980).

35   G. Toulouse, *Commun. Phys.* **2**, 115 (1977).

36   D. L. Stein and P. W. Anderson, *Proc. Natl Acad. Sci. USA* **81**, 1751 (1984).

37   J. D. Bryngelson and P. G. Wolynes, *Proc. Natl Acad. Sci. USA* **84**, 7524 (1987).

38   A. M. Gutin and E. I. Shakhnovich, *J. Chem. Phys.* **98**, 8174 (1993).

39   B. Derrida and L. Peliti, *Bull. Math. Biol.* **53**, 355 (1991).

40   S. Kauffman and S. Levin, *J. Theor. Biol.* **128**, 11 (1987).

41   H. Kitano, *Nature Rev.* **5**, 826 (2004).

42   G. Weisbuch, *J. Theor. Biol.* **143**, 507 (1990).

43   B. Drossel, *Adv. Phys.* **50**, 209 (2001).

44   J. Sun, D. J. Earl, and M. W. Deem, *Modern Phys. Lett. B* **20**, 63 (2006).

45   A. S. Perelson and C. A. Macken, *Proc. Natl Acad. Sci. USA* **92**, 9657 (1995).

46   S. A. Kauffman and W. G. MacReady, *J. Theor. Biol.* **173**, 427 (1995).

47   T. Tan, L. D. Bogarad, and M. W. Deem, *J. Mol. Evol.* **59**, 385 (2004).

48   G. M. Edelman, *Sci. Amer.* **223**, 34 (1970).

49   R. R. Porter, *Science* **180**, 713 (1973).

50   D. L. French, R. Laskov, and M. D. Scharff, *Science* **244**, 1152 (1989).

51   J.-M. Park and M. W. Deem, *Physica A* **341**, 455 (2004).

52   R. M. Zinkernagel and H. Hengartner, *Science* **293**, 251 (2001).

53   A. W. Goldrath and M. J. Bevan, *Nature* **402**, 255 (1999).

54   P. A. van der Merwe and S. J. Davis, *Annu. Rev. Immunol.* **21**, 659 (2003).

55   H. L. Robinson and R. R. Amara, *Nature Med.* **11**, S25 (2005).

56   Q. Ge, A. Bai, B. Jones, H. N. Eisen, and J. Chen, *Immunology* **101**, 3041 (2004).

57   R. M. Kedl, J. W. Kappler, and P. Marrack, *Curr. Opin. Immunol.* **15**, 120 (2003).

58   M. A. Purbhoo, D. J. Irvine, J. B. Huppa, and M. M. Davis, *Nature. Immunol.* **5**, 524 (2004).

59   C. Kesmir, J. A. M. Borghans, and R. J. de Boer, *et al. Science* **288**, 1135 (2000).

60   T. P. Arstila, A. Casrouge, V. Baron, *et al.*, *Science* **286**, 958 (1999).

61   M. J. Miller, S. H. Wei, I. Parker, and M. D. Cahalan, *Science* **296**, 1869 (2002).

62   T. R. Mempel, S. E. Henrickson, and U. H. von Adrian, *Nature* **427**, 154 (2004).

63   A. T. Stock, S. N. Mueller, A. L. van Lint, W. R. Heath, and F. R. Carbone, *J. Immunol.* **18**, 2241 (2004).

64   J. N. Blattman, D. J. D. Sourdive, K. Murali-Krishna, R. Ahmed, and J. D. Altman, *J. Immunol.* **165**, 6081 (2000).

65   D. J. D. Sourdive, K. Murali-Krishna, J. D. Altman, *et al.*, *J. Exp. Med.* **188**, 71 (1998).

66   A. Mathew, I. Kurane, A. L. Rothman, *et al.*, *J. Clin. Invest.* **98**, 1684 (1996).

67   J. Zivny, I. Kurane, A. M. Leporati, *et al.*, *J. Exp. Med.* **182**, 853 (1995).

68   D. Gray, *Annu. Rev. Immunol.* **11**, 49 (1993).

69   F. L. Black and L. Rosen, *J. Immunol.* **88**, 725 (1962).

70   J. Sprent, *Curr. Opin. Immunol.* **5**, 433 (1993).

71   S. Fazekas de St. Groth and R. G. Webster, *J. Exp. Med.* **124**, 331 (1966).

72   S. Fazekas de St. Groth and R. G. Webster, *J. Exp. Med.* **124**, 347 (1966).

73   C. Berek and C. Milstein, *Immunol. Rev.* **96**, 23 (1987).

74   D. J. Smith, S. Forrest, D. H. Ackley, and A. S. Perelson, *Proc. Natl Acad. Sci. USA* **96**, 14 001 (1999).

75   G. M. Air, M. C. Els, L. E. Brown, W. G. Laver, and R. G. Webster, *Virology* **145**, 237 (1985).

76   W. M. Fitch, J. M. Leiter, X. Li, and P. Palese, *Proc. Natl Acad. Sci. USA* **88**, 4270 (1991).

77   R. M. Bush, C. A. Bender, K. Subbarao, N. J. Cox, and W. M. Fitch, *Science* **286**, 1921 (1999).

78   J. B. Plotkin and J. Dushoff, *Proc. Natl Acad. Sci. USA* **100**, 7152 (2003).

79   W. M. Fitch, R. M. Butch, C. A. Bender, K. Subbarao, and N. J. Cox, *Heredity* **91**, 183 (2000).

80   C. Macken, H. Lu, J. Goodman, and L. Boykin, in *Options for the Control of Influenza IV*, ed. by A. D. M. E. Osterhaus, N. Cox, and A. W. Hampson (New York, Elsevier Science, 2001), hemagglutinin H3 epitope structural mapping, www.flu.lanl.gov/.

81   J. W. Smith and R. Pollard, *J. Hyg.* **83**, 157 (1979).

82   W. A. Keitel, T. R. Cate, and R. B. Couch, *Am. J. Epidemiol.* **127**, 353 (1988).

83   W. A. Keitel, T. R. Cate, R. B. Couch, L. L. Huggins, and K. R. Hess, *Vaccine* **15**, 1114 (1997).

84   V. Demicheli, D. Rivetti, J. Deeks, and T. Jefferson, *Cochrane Database Syst. Rev.* **3**, CD001269 (2004).

85   K. M. Edwards, W. D. Dupont, M. K. Westrich, *et al.*, *J. Infect. Dis.* **169**, 68 (1994).

86   D. S. Campbell and M. H. Rumley, *J. Occup. Environ. Med.* **39**, 408 (1997).

87   K. L. Nichol, A. Lind, K. L. Margolis, *et al.*, *New Engl. J. Med.* **333**, 889 (1995).

88   I. Grotto, Y. Mandel, M. S. Green, *et al.*, *Clin. Infect. Dis.* **26**, 913 (1998).

89   M. L. Clements, R. F. Betts, E. L. Tierney, and B. R. Murphy, *J. Clin. Microbiol.* **23**, 73 (1986).

90   C. B. Bridges, W. W. Thompson, M. I. Meltzer, *et al.*, *J. Am. Med. Assoc.* **284**, 1655 (2000).

91   M. A. Mixeu, G. N. Vespa, E. Forleo-Neto, J. Toniolo-Neto, and P. M. Alves, *Aviat. Space Environ. Med.* **73**, 876 (2002).

92   J. L. Millot, M. Aymard, and A. Bardol, *Occup. Med.-Oxford* **52**, 281 (2002).

93   N. Kawai, H. Ikematsu, N. Iwaki, *et al.*, *Vaccine* **21**, 4507 (2003).

94   R. T. Lester, A. McGeer, G. Tomlinson, and A. S. Detsky, *Infect. Cont. Hosp. Epidemiol.* **24**, 839 (2003).

95   S. Dolan, A. C. Nyquist, D. Ondrejka, *et al.*, *Centers for Disease Control and Prevention Morbidity and Mortality Weekly Report* **53**, 8 (2004).

96   M. S. Lee and J. S. E. Chen, *Emerg. Infect. Dis.* **10**, 1385 (2004).

97   N. Cox, A. Balish, L. Brammer, *et al.*, in *Information for the Vaccines and Related Biological Products Advisory Committee, CBER, FDA, WHO Collaborating Center for Surveillance Epidemiology and Control of Influenza* (Atlanta, GA, Centers for Disease Control, 2003).

98   G. W. Both, M. J. Sleigh, N. J. Cox, and A. P. Kendal, *J. Virol.* **48**, 52 (1983).

99   J. S. Ellis, P. Chakraverty, and J. P. Clewley, *Arch. Virol.* **140**, 1889 (1995).

100   M. T. Coiras, J. C. Aguilar, M. Galiano, *et al.*, *Arch. Virol.* **146**, 2133 (2001).

101   *WHO Weekly Epidemiological Record* **63**, 57 (1988).

102   A. V. Pontoriero, E. G. Baumeister, A. M. Campos, *et al.*, *Pan Am. J. Publ. Health* **9**, 246 (2001).

103   Information for FDA vaccine advisory panel meeting, 1997, Atlanta, GA, Centers for Disease Control.

104   R. A. Goldsby, T. J. Kindt, B. A. Osborne, and J. Kuby, *Immunobiology*, 5th edn. (New York, W. H. Freeman, 2002).

105   M. Ney-Nifle and A. P. Young, *J. Phys. A* **30**, 5311 (1997).

106   V. Azcoiti, E. Follana, and F. Ritort, *J. Phys. A* **28**, 3863 (1995).

107   P. D. Holler, L. K. Chlewicki, and D. M. Kranz, *Nature Immunol.* **4**, 55 (2003).

108   E. T. Munoz and M. W. Deem, *Vaccine* **23**, 1142 (2004).

109   N. A. Campbell and J. B. Reece, *Biology*, 7th edn. (San Francisco, CA, Benjamin Cummings, 2004).

110   J. K. Percus, O. E. Percus, and A. S. Perelson, *Proc. Natl Acad. Sci. USA* **90**, 1691 (1993).

111   V. Detours, H. Bersini, J. Stewart, and F. Varela, *J. Theor. Biol.* **170**, 401 (1994).

112   J. Sulzer, L. van Hemmen, A. U. Newmann, and U. Behn, *Bull. Math. Biol.* **55**, 1133 (1993).

113   T. Tan, D. Frenkel, V. Gupta, and M. W. Deem, *Physica A* **350**, 52 (2005).

114   M. Larché and D. C. Wraith, *Nature Med.* **11**, S69 (2005).

115   A. Crameri, S. A. Raillard, E. Bermudez, and W. P. C. Stemmer, *Nature* **391**, 288 (1998).

116   M. Ostermeier, J. H. Shim, and S. J. Benkovic, *Nature Biotech.* **17**, 1205 (1999).

117   W. J. Netzer and F. U. Hartl, *Nature* **388**, 343 (1997).

118   D. Kaplan, I. Ferrari, P. L. Bergami, *et al.*, *Proc. Natl Acad. Sci. USA* **94**, 10301 (1997).

119   M. Leirisalo-Repo, *Curr. Opin. Rheumatol.* **17**, 433 (2005).

120   S. M. Carty, N. Snowden, and A. J. Silman, *J. Rheumatol.* **30**, 425 (2003).

121   J. Mongkolsapaya, W. Dejnirattisai, X. Xu, *et al.*, *Nature Med.* **9**, 921 (2003).

122   A. L. Rothman, *J. Clin. Invest.* **113**, 946 (2004).

123   P. Klenerman and R. M. Zinkernagel, *Nature* **394**, 482 (1998).

124   N. Sangkawibha, S. Rojanasuphot, S. Ahandrik, *et al.*, *Am. J. Epidemiol.* **120**, 653 (1984).

125   R. M. Welsh and A. L. Rothman, *Nature Med.* **9**, 820 (2003).

126   A. L. Rothman, N. Kanesa-thasan, K. West, *et al.*, *Vaccine* **19**, 4694 (2001).

127   T. Dharakul, I. Kurane, N. Bhamarapravati, *et al.*, *J. Infect. Dis.* **170**, 27 (1994).

128   P. L. Nara and R. Garrity, *Vaccine* **16**, 1780 (1998).

129   J. W. Yewdell and J. R. Bennink, *Annu. Rev. Immunol.* **17**, 51 (1999).

130   A. A. Freitas and B. Rocha, *Annu. Rev. Immunol.* **18**, 83 (2000).

131   M. W. Deem, *AIChE J.* **51**, 3086 (2005).

132   H. Schreiber, T. H. Wu, J. Nachman, and W. M. Kast, *Semin. Cancer. Biol.* **12**, 25 (2002).

133   A. Martin-Fontecha, S. Sebastiani, U. E. Hopken, *et al.*, *J. Exp. Med.* **198**, 615 (2003).

134   H. Hon and J. Jacob, *Immunologic Res.* **29**, 69 (2004).

135   R. G. van der Most, A. Sette, C. Oseroff, *et al.*, *J. Immunol.* **157**, 5543 (1996).

136   G. A. Cole, T. L. Hogg, M. A. Coppola, and D. L. Woodland, *J. Immunol.* **158**, 4301 (1997).

137   A. Makki, G. Weidt, N. E. Blachere, L. Lefrancois, and P. K. Srivastava, *Cancer Immunity* **2**, 4 (2002).

138   A. C. Spaulding, I. Kurane, F. A. Ennis, and A. L. Rothman, *J. Virol.* **73**, 398 (1999).

139   D. J. Gubler, *Nature Med.* **10**, 129 (2004).

140   R. L. Fournier, *Basic Transport Phenomena in Biomedical Engineering*, 2nd edn. (Philadelphia, PA, Taylor & Francis, 2007).

Index